Markets, Resources
and the Environment

National Priorities Project 1991
Markets, Resources and the Environment

Edited by
Alan Moran, Andrew Chisholm
and Michael Porter

including contributions by
Peter Ackroyd, Terry Anderson, John Fogarty,
John Freebairn, Peter Hartley, Rodney Hide,
Anthony Quin and Ian Wills

ALLEN & UNWIN
in association with
National Priorities Project
Tasman Institute

© Moran, Chisholm and Porter, 1991
This book is copyright under the Berne Convention. No
reproduction without permission. All rights reserved.

First published in 1991
Allen & Unwin Pty Ltd
8 Napier Street, North Sydney 2059, Australia

National Library of Australia
Cataloguing-in-Publication entry:

Markets, resources and the environment.

Bibliography.
ISBN 1 86373 025 7.
1. Economic development – Environmental aspects.
2. Environmental policy – Economic aspects –
Australia. I. Moran. Alan J. II. Chisholm, A.
(Andrew), 1959 –   III. Porter, Michael G. (Michael
Glenthorne), 1943 –   . IV. National Priorities
Project (Australia). V. Tasman Institute.

333.7

Set in 11pt Times in Ventura by Tasman Institute, Melbourne
Printed and bound in Australia by
The Book Printer, Victoria

Contents

Preface vii

Contributors xi

1 Economics and the environment — the Australian debate 1

2 A framework for policy 16

3 Managing the environment: an historical perspective 46

4 Land degradation 76

5 Wood, wildlife and wilderness: managing Australia's native timber forests 102

6 Mining and the environment 134

7 Markets and sewerage 161

8 A case study—establishing property rights to Chatham Islands' abalone (paua) 188

9 Air pollution 207

10 The enhanced greenhouse effect 235

Glossary 266

Abreviations 274

References 275

Appendix 289

Preface

This volume is the fourth in the National Priorities Project (NPP) series of major strategic reviews of fundamental issues affecting the performance of the Australian economy. The first two volumes, *Spending and Taxing*, and *Spending and Taxing II*, dealt with major taxation, spending and associated microeconomic reforms which research suggests as necessary in order to lift Australia's economic performance. The 1989 volume, *Savings and Productivity*, focussed on strategies capable of achieving a substantial lift in Australia's productivity and savings rate, so laying the foundations for improved international competitiveness and enhanced living standards.

This 1991 volume of the NPP responds to the environmental debate and argues that the pursuit of efficient economic growth remains the key to both environmental and economic policy design. Properly defined measures of national income can embrace the environmental outcomes at the heart of our quality of life. The challenge is to bring environmental considerations explicitly into the economic calculus, through assignment of property rights and obligations, and if necessary, 'polluter pays' charges and taxes.

The alleged conflict between economic growth, correctly defined, and environmental goals is found to arise only where entitlements and liability are poorly defined. In order to bring economic growth and the environment into harmony, this volume seeks, therefore, to unleash market forces and improved incentives so as to facilitate the efficient achievement of economic and environmental goals. We argue that research clearly shows the superiority of property rights and incentive based approaches over those control and command mechanisms characteristic of centrally planned economies.

At a time when Eastern European countries are starting the arduous transition from command and control structures towards a market system based on property rights and the rule of law, it would be ironic indeed if Australia and other countries responded to the environment movement and pollution issues by shifting to the sort of control systems which have impoverished the socialist bloc, and in the process, created in Eastern Europe some of the world's worst environmental disasters.

Tasman Institute—Project Manager

Tasman Institute is delighted to have been commissioned by the NPP to produce this volume. Tasman's pioneer research program is on the theme of 'Markets and the Environment' and production of the volume for the NPP was a timely opportunity. While the sponsors of the 1991 NPP volume are listed at the back of the book, we also wish to acknowledge the Tasman Institute project which has enabled us to assemble a distinguished array of national and international experts on economic and environmental issues.

Authorship

Editorial responsibility for this volume has been taken by Alan Moran, Research Director, Tasman Institute, Andrew Chisholm, Research Fellow, Tasman Institute and myself. While the editors take responsibility for the substance of this volume, the principal authors of each chapter were:

Chapter 1 'Economics and the environment—the Australian debate' (Michael Porter);

Chapter 2 'A framework for policy' (Alan Moran, Andrew Chisholm, Peter Hartley and Michael Porter);

Chapter 3 'Managing the environment: an historical perspective' (John Fogarty);

Chapter 4 'Land degradation' (John Freebairn);

Chapter 5 'Wood, wildlife and wilderness: managing Australia's native timber forests (Andrew Chisholm and Terry Anderson);

Chapter 6 'Mining and the environment' (Alan Moran);

Chapter 7 'Markets and sewerage' (Ian Wills and Anthony Quin);

Chapter 8 'A case study—establishing property rights to Chatham Islands' abalone (paua) (Peter Ackroyd and Rodney Hide);

Chapter 9 'Air pollution' (Alan Moran);

Chapter 10 'The enhanced greenhouse effect' (Alan Moran and Andrew Chisholm).

Particular acknowledgements go to Visiting Scholars at Tasman Institute, in particular, Terry Anderson, Peter Hartley and Mikhail Bernstam, Carol MacSporran of the CSIRO, for her assistance with the references on the greenhouse chapter, and Elizabeth Prior, for editorial assistance.

We are especially thankful to the patient and professional assistance of the Managing Editor, Hazel Ramsden, who administered the production of the book. We also thank Ursula Canty, Anne Bluett, Cornelis Romein and Julie Robertson for producing the document within very tight deadlines.

Michael Porter
Executive Director
Tasman Institute
January 1991

Contributors

Peter Ackroyd is Research Officer at the Centre For Resource Management, Lincoln University, New Zealand.

Terry Anderson is Professor of Economics at Montana State University and Senior Associate, the Political Economy Research Center, Bozeman, Montana.

Andrew Chisholm is Research Fellow at Tasman Institute.

John Fogarty is a consultant historian.

John Freebairn is Professor of Economics and Chairman, Department of Economics, Monash University.

Peter Hartley is Professor of Economics, Rice University, USA, and Visiting Fellow at Tasman Institute.

Rodney Hide is Lecturer, The Economics and Marketing Department, Lincoln University, New Zealand.

Alan Moran is Research Director at Tasman Institute.

Michael Porter is Executive Director at Tasman Institute.

Anthony Quin is a Tasman Fellow at Tasman Institute.

Ian Wills is Senior Lecturer, Department of Economics, Monash University.

1 Economics and the environment — the Australian debate

Executive summary

The environmental debate in Australia and overseas is cast in terms which often suggest that economic growth and environmental objectives are in conflict. Cross-country comparisons suggest, to the contrary, that economic growth and environmental enhancement go together. Increased income enables us to afford greater spending on environmental amenities. It is also clear that Eastern European and other poor countries have far inferior environmental conditions—yet many, spurred on by the views of Dr Suzuki and other media commentators, continue to see economic growth as a major environmental problem.

Although trade-offs must inevitably be made, analysis of economic development in Australia also tells us that economic and ecological goals are mutually supporting. Income growth and environmental care go hand-in-hand. This is particularly so when sensible economic rules apply, and if valuable amenities are vested in owners with a 'duty of care'. The history of progress in Australia is very much the history of the use, re-development and re-use of a very wide range of natural resources, notably land, minerals and our oceans and other water resources. In the case of minerals, it is clearly possible to extract value, to generate a wide range of manufacturing and service activities, and subsequently to restore the land to attractive and even superior forms.

Australia has, despite its labour market and regulatory disadvantages, a competitive edge in a wide range of resource and land based projects. The resources sector, including agriculture and mining, accounted for between 65% and 73% of exports of goods

2 Markets, Resources and the Environment

and services over the 1980s. These primary sectors generated directly 9%–10% of GDP at factor cost over the 1980s. These areas of competitive strength, in turn, sustain manufacturing and service activities on a large scale and generate a major share of community income.

Australia also has massive tourism potential, as we and the rest of the world seek to enjoy our abundant natural endowments. To capitalise on increased tourism often means engaging in development, but it can also mean preservation of what we have. If, despite these opportunities, there is a surge of opinion in favour of restraining growth and turning off major development projects, then Australia has every chance of becoming a nation unable to afford the good things in life, including clean air and water, and a wide range of attractive environmental amenities. Policies based on limiting economic growth are a real danger to sustaining and enriching the Australian way of life. Australia needs increased investment, higher savings, and an expansion of the quality and quantity of capital deployed.

Capitalism, property rights and markets

The National Priorities Project argues that economic growth is part of the solution to the environmental challenge. Growth enables us to afford improved technologies to deal with water, air and congestion problems. The fundamentals of capitalism, such as the careful designation of property rights and obligations, the enforcement of common law, and the facilitation by government of trading in emissions quotas and so forth, are the essence of efficient solutions to the environmental challenges. The major failures in the environmental area, as seen in this book, stem not from the market or from economic growth, but from the failure, often of government, to facilitate market transactions over key parts of our economy in which environmental 'bads' are produced.

As an example, emissions and effluent can, under certain designations of property rights, be well described as invasions of private property—as people are tipping their garbage, if you like, into others' spaces. The difficulty is that property rights over the water, air and land which is invaded by unwelcome pollution, are usually poorly defined. Often this is because the cost of defining and monitoring makes a true vesting of such rights infeasible or technically difficult. There may also be a lack of common law access to compensation, such that the well meaning actions of free agents in the market place produce inferior outcomes to those which would arise with properly designated trading in environmental 'bads'.

The challenge for government, for bureaucrats and for the community at large, is to devise 'rules of the game' to minimise these areas where it is not possible for individuals to express their own preference judgements. Individuals acting privately or in groups can best enhance their future and present opportunities. Self interest, rather than 'command and control' strategies, are more reliable ways of generating improvements, whether economic or environmental. The literature on privatisation, and on the performance of capitalist versus collectivist economic systems in recent decades, tells us that well defined property rights, clear intelligible and properly enforced laws, and monitoring of behaviour, are the logical and central function of government. Our prosperity hinges quite crucially on a system in which property is privately owned or managed within the common law, and subject to the rules and covenants laid down by government. It is, therefore, unsurprising that the most polluting areas of the world happen to be in the communist countries where property rights are poorly defined.

In Australia, as in other advanced industrial countries, the areas most problematic in terms of environmental pollution include vehicle emissions, toxic waste, pollution of oceans through sewage, and so forth. These are areas in which property rights are being abused—or where entitlement is difficult to vest. Pollution is often a problem of failure to construct markets based on private property, rather than a failure of markets. The problem may be due to a failure of government to install proper incentive systems, or an unwillingness to create the basis for markets and trading in environmental products, or bi-products. Designating water, fishing, air, emissions, and noise rights, and then allowing trading in these rights is a logical function for government, often at the local level.

4   Markets, Resources and the Environment

A further problem is that governments have, courtesy of taxpayers, rather deep pockets. Hence, the state enterprises such as those in transport, electricity, waste management and so forth, can be funded and their poor performance hidden for decades by the capacity of governments to over-charge for other services, or use taxes to fund deficits. There is rarely the incentive in such organisations to do what is most efficient, but rather an incentive to assist interest groups who happen to have political or other forms of power. Moreover, where government agencies violate environmental standards, it is unlikely that government will ever prosecute an offender if the offender is another arm of government. However, with properly defined property rights, with private agencies taking responsibility for water, land, ocean facilities, and so forth, and with legally accountable boards and chief executives, there are incentive mechanisms for holding accountable the polluters who violate standards and covenants.

Technology is also bringing the potential for additional rights to be brought within the market system. For example, recent research suggests that car emissions such as carbon monoxide and SO_2 can be monitored at minuscule cost per car using modern day technology involving computers and scanners. The location of cars can also be monitored electronically, at very small unit cost, thereby making it possible for 'fast lane clubs', and for congestion taxes to be imposed on the owners of cars causing the problem. Taxes can, with appropriate technology, also be applied according to the volume and nature of the emissions.

Private ownership and market approaches can, then, be a powerful force enhancing environmental performance, once rules have been set. Capitalism is naturally a conserver of resources, and an enemy of waste. It is not therefore a question of less or more market activity, nor of big government or small government. It is not a debate between privatisation and state ownership. Rather, it is a debate about government doing what it is comparatively good at—defining and enforcing the rules—and withdrawing from what it is bad at—managing and owning enterprises—so that the private sector can profit from achieving goals within the government set rules.

Ecologically sustainable development

It is hard to imagine that anyone in Australia is in favour of *un*sustainable development. Yet there is now a major debate, replete with Prime Ministerial working parties, on a Government discussion paper entitled *Ecologically Sustainable Development*. The soothing verbal cocktail of 'ecologically sustainable development' is at best a form of motherhood, and at worst a dangerous brew. In the hands of an anti-development brigade, 'sustainability', to use the shorthand, could be used to stop or slow down key resource intensive projects, which are capable both of raising our standard of living and enabling us to afford enhanced environmental standards.

As an example of the ideas of this book, the debate on 'ecologically sustainable development' is seen as an opportunity to demonstrate that economic growth, properly defined, and well constructed market systems, are compatible with the goal of providing future generations with resources and opportunities even more valuable than those enjoyed by current generations.

What we do not endorse, and what we fear may be endorsed by some advocates of 'sustainability', is a policy of preventing any use of particular resources on the grounds that they are scarce. All resources are physically scarce, and many non-renewable resources are clearly in limited physical supply. But economic history has demonstrated that it is our intellectual resources, our human ingenuity, our capacity to create substitutes and alternative ways of doing things, which explains not merely that exhaustion of natural resources is not a particularly critical problem, but on the contrary, that the relative price of most of these natural resource based commodities has in fact fallen over the last century.

The real price of primary commodities fell on average by 0.5% per annum, from 1900 to 1986 (see Grilli and Yang, 1987) thanks to technical innovations and the discovery of widespread substitution possibilities. The Club of Rome, and other doomsdayers have been proven wrong by the capacity of intellectual resources—and resulting technology—to create effectively infinite substitution possibilities out of finite physical or primary resources. On current trends of mineral resource depletion, given the natural occurrence of minerals, there is no metal which would need to be wound back, so long as relative prices are allowed to rise, in some cases by a factor of four.

Contrary to the popular wisdom espoused by many of the less rational environmentalists, our scarce minerals have been falling in price for centuries, and even oil has fallen in price relative to manufactures apart from times of politically induced shortages. Designated oil reserves always seem to be scheduled to fall a few decades hence, yet three decades later it turns out that yet further discoveries have taken place, making those forecasts turn out to be quite wrong.

What happens is that expected scarcities and higher relative prices subsequently generate exploration, new technologies and so an upward revision in the supply of substitutes for exhaustible resources. Temporary price rises create incentives to find substitutes, to explore new areas, and generally to economise on the use of a particular resource. When it comes to energy resources in general, and oil in particular, the abundance of coal in Australia and the world, the capacity to use solar energy, the availability of nuclear alternatives and many other technological opportunities will continue, we suggest, to ensure that the supply of oil is far from being a fundamental problem. While it is true that liquid fuels sell at a premium because of their portability, and while short term shortages, due to the crisis of the moment, can create temporary problems, this is more a matter of proper inventory management, and of international political and economic strategy than it is of fundamental resource scarcity.

Looking after future generations

A healthy and wealthy Australia will also choose to reserve for the future similar, but not identical, endowments to those we have inherited. Such natural human tendencies ensure that future generations have at least as many options as those that exist today. However, we should not cast economic development in concrete—nor seek to impose a fixed plan or seek to further rigidify the Australian economy. The essence of economic progress is that opportunities and tastes change, ideas develop and technology advances. Men and women are dynamic learning-oriented individuals, particularly when incentives and opportunities are in place. What is optimal today may be an irrelevant alternative tomorrow, as mankind finds new and better ways of achieving an improved quality of life.

One sure thing about our future is that we do not know it in detail, but we do know that we need incentives to discover and adapt to new information, and not an obligation to conform to predetermined stereotypes. What we wish to achieve is a preservation of options, and the key to this is the sustained and efficient development of our human, physical and intellectual resources.

Our concern for future generations suggests, at least to this set of authors, that labour market reform, tax reform, monetary discipline and a smaller role for government in enterprises and regulation, are the real key to future prosperity. And most importantly, a more prosperous Australia will also be able to afford a higher level of environmental amenities, superior devices for monitoring and controlling emissions, and will have the capacity, for example, to convert waste water into fertiliser or process further out to sea, rather than polluting our beaches with effluent.

The question of risk

To some, the possibility of greenhouse effects, holes in the ozone layer, depletion of oil resources and so forth, all mean that we are placing our grandchildren at risk in continuing down the current development path. But the risk analysis goes both ways. Any strategy which would stop Australia's major mineral and agricultural projects, hold up pulp and paper processing, prevent logging in a sustainable and economically efficient fashion, is more likely to so lower our standards of living. Such a strategy would limit the economic opportunities for our grandchildren, so that Australia would indeed become the poor white nation of the Pacific, as well as a country with lower environmental standards. Furthermore, an Australia racked by social division, and an Australia doing poorly relative to other countries in the world, is unlikely to be a country free of major social and political risks.

8   Markets, Resources and the Environment

The chapters in brief

In what follows we summarise briefly the issues at the heart of the chapters which follow. Readers with more time, may prefer to proceed at this point to the chapters themselves, rather than suffer the over simplifications inherent in these summaries of summaries!

Chapter 2 brings together the key economic issues relating to environmental issues, and this constitutes an overview of the analytical framework which underlies the book. These issues have been summarised above, and generally suggest that a *property rights* approach is, in general, preferred to *'command and control'* strategies if the twin objectives of economic growth and environmental enhancement are to be placed in a reasonable balance.

Chapter 3 sets out the broad historical context, including the evolution of our settlement policies, which led to the clearing of land in Australia, under varying patterns of ownership and regulation of land use by government. We note that the enhancement of the lives of all Australians has clearly involved a dramatic transformation of the Australian landscape. Our city and rural areas, once wild scrub, became quite magnificent places in which to live. While mistakes were made, and while land degradation and salination have been a problem in specific instances, usually these difficulties resulted from inadequate, wrong, or costly information. Typically the problem was the nature of the incentives at the time, rather than any intrinsic tendency amongst the early settlers and farmers, for example, to act in an ecologically unsound manner.

Mineral development has also been a major source of wealth in Australia, with a relatively minor impact on the environment. This is not to say that particular mines have not occasionally had negative impacts. The issue ultimately relates to property rights, incentives and the ways in which the wealth generated is applied to restoration of mine sites and to other activities which enhance our lives. There is nothing intrinsic in mining which harms the environment—on the contrary. History has clearly demonstrated the capacity of mineral and agricultural activity to dramatically increase Australian living standards, broadly defined.

A century ago, when Australia very much rode on the sheep's back, with major assistance from mineral activities, we were arguably the richest country in the world per capita, and could no doubt afford environmental standards second to none. While subsequently our tariff, taxation, labour market, and other policies have caused Australia to fall behind in the economic growth stakes, and have marginally reduced our capacity to afford high environmental standards, such as clean beaches, we nevertheless continue to be a country better able than most to afford world class environmental amenities.

Land degradation

The clearing of forests to create pastures and urban developments on former agricultural land, are both examples of the sort of environmental transformation which has, by some, been labelled 'land degradation'. We argue, to the contrary, that when faced with the relevant information regarding erosion, acidification and soil fertility, the Australian farm sector has had a spectacularly good record, both in productivity terms and in maintaining an attractive rustic environment. By and large, the incentives facing farmers encourage them to conserve and enhance the land. While it is true that some soil erosion has taken place, typically the costs of erosion have been borne, or at least shared, by the individual farmer. It is nomadic farming, with no rights to ownership of land, which tends to lead to rape of the land, and create erosion. Nomads extract value and move onto the next field, rather than clearing, fertilising, maintaining and indeed, enhancing the productive capacity of the land.

Fundamental to reducing land degradation in Australia is the proper designation of property rights to land, with squatting, and then leasehold and freehold land creating the incentives necessary for sustainable development.

When it comes to irrigation, the key, again, is the careful designation of property rights, in this case for water usage rights, with failure properly to designate entitlements having the capacity both to waste water and to create salinity problems through excessive extraction of water from the rivers. A large part of the difficulty, in the case of irrigation, stems from the failure of government to allow tradeable quotas in water rights. In our view while land degradation and irrigation have been problem areas, the problem has been too little, rather than too great, a use of markets.

Wood, wildlife and wilderness

Just as poorly designated land and water rights can create problems of land degradation and salinity, so too can political ownership of forests and recreational areas inadvertently create patterns of logging, use and abuse of recreational facilities which are far from satisfactory. It is the burden of Chapter 5 of this book that allowing the owners of forest resources to include private parties, including environmental groups, possibly subject to government covenants and other restrictions, will enable more appropriate trade offs between competitive forests uses. These would include timber and use of forests for the environmental amenities provided.

All state governments in Australia have, from late in the 19th Century, pursued a policy of public ownership and management of native forests. Originally driven by fears of timber shortages, 73% of Australia's native forests are now managed by state departments and forestry commissions, yet these enterprises are typically returning inadequate value to their owners, by logging and pricing in ways which do not reflect scarcity values.

These commercial failures of the state forestry enterprises are also associated with perceived failures on the environmental front. In the pursuit off their environmental objectives, environmental groups have blockaded logging operations and demanded complete withdrawal of timber interests from state forests.

We argue that the goal of forest management should not be an 'all or nothing' approach, but rather aim at a balance of competitive uses of forests. This balance can be achieved by allowing community valuations of timber to be reflected fully in timber prices, and for mixed leisure, mineral and timber uses of forest lands, possibly subject to covenants reflecting environmental objectives laid down by Parliament.

The strategies and policies evaluated in this Chapter include:
- traditional 'sustained yield—even flow management', as commonly practised by the forests services;
- economic management for timber production alone;
- 'scientific' multiple use management with public ownership;
- vesting of private rights to forests in conservation groups and the timber industry, and the use of market based prices.

We argue that research demonstrates that proper economic management of privately owned forests, subject to government covenants, would tend to lead to fewer forests being logged, with the forests being managed more productively, both in terms of timber use and provision of environmental amenities.

Mining and the environment

Australia is a world leader in energy and resources, with minerals and processed mining products accounting for roughly 50% of Australia's exports. Australian mining industry output has grown eight-fold over the last quarter of a century and accounts for a good deal of the standard of living of Australians.

Whether it be mineral sands, uranium, diamonds, coal or iron, mining neither creates nor destroys resources, but it does transform what is often barren territory into commodities of considerable value. While there are mining sites, like farm areas, which have been less than perfectly restored or maintained, proper specification of rights, obligations and 'the duty of care' is capable of giving us both the economic fruits of mineral development and the enhanced environment which can flow naturally from a wealthier and healthier community.

Markets and sewerage

The sewerage industry has a written down asset value of $18 billion (Industry Commission, 1990) making it one of our largest industry sectors. The capital sums involved in providing sewerage services together with actual and potential environmental problems associated with the industry, make it imperative that this be one of the more efficient and well managed industries in the country. Regrettably, the 'waste water industry', has, in conjunction with the water supply industry, been a state monopoly, with little attempt to price services or to allow private sector competition in waste processing. That said, natural monopoly and externality arguments make a case for government involvement—the question is how to achieve private sector efficiencies without allowing the abuse of the monopoly powers that might arise owing to the existence of a natural monopoly over the pipes and associated infrastructure.

There have been a number of innovations in recent years, for example, in the US private companies have been installing waste treatment plants in competition with the state sector. In France, while municipalities have continued to own the pipe networks, the provision of water and waste water services has been through competitive franchising, with the terms and conditions of the franchise being designed to achieve both competitive outcomes and maintenance networks.

In the UK, in 1988, the water and waste water industry was privatised with assets being sold into separate PLCs which, nevertheless, have geographical monopolies based on the river basins. Attempts to protect consumers against monopoly power have led to very detailed licensing arrangements and, at the time of writing, there is a question mark as to whether private monopolies and centralised regulation are delivering an improved quality of service.

Fishing and property rights

A good example of the application of property rights concepts in the environmental debate relates to the problem of 'over-fishing'. Over-fishing is not intrinsically a problem of private enterprise, but a consequence of common ownership—or more precisely lack of ownership—of the fish resource. In this chapter we explore the notion that government should issue and enforce ceiling quotas of fishing rights, with fishermen competing to purchase quotas, which are set such that the size of the resource is sustained, not undermined.

Our particular case study comes from the Chatham Islands in NZ. These islands, like other fishing areas, have, over the last century, experienced spectacular booms and busts in relation to seals, whales, cod, and most notoriously, rock lobster. Absence of defined rights to fish in these and other areas causes rapid depletion of the resource. In 1987 abalone (paua) fishermen were presented with individual transferable quotas (ITQs) that allow the taking of a specific tonnage of fish per annum. What the study reveals is that the careful designation of rights to fish facilitated profitable management of a valuable resource, while sustaining the resource.

Air pollution

Casual reading of the popular press and much localised experience, has convinced many that air pollution has been a general and growing problem in Western countries. However, there is much evidence of the successful resolution of air pollution problems. Today the air of most developed countries is much cleaner than it was a century ago, not withstanding traffic growth and increased energy consumption. Whereas influenza, pneumonia and tuberculosis—all largely attributable to pollution—accounted for about 25% of the deaths at the end of the 19th Century, they now account for less than 5% in a population in which life expectancy has increased by 50%.

Air pollution in Australia has also decreased to a surprising degree in many areas over recent years, for example, sulphur dioxide levels have trended downwards and in 1988 were less than 50% of the government designated maximum acceptable peak. While emissions problems are far from trivial, we suggest in Chapter 9, that new technologies responding to tradeable emissions quotas, registration charges which reflect emissions and other market-based charges, have a capacity significantly to reduce environmental pollution.

The abatement of urban air pollution levels has been achieved by *'command and control'* regulation. Where markets do not automatically equilibrate supply and demand because of monitoring difficulties, total permitted supply could be specified by a government authority, with companies acquiring and trading quotas to the limited emissions. Such quasi-market approaches will pay dividends when applied to some sources; however continuation of 'command and control' approaches seems to be inevitable in the case of domestic and, perhaps, automotive emissions. In a strict sense, therefore, the achievement of efficiency largely turns on the *nature* of the regulation. In Chapter 9 we argue that if market mechanisms are employed to allow polluters flexibility in meeting the levels desired, then we can achieve the same emissions standards at a reduced cost.

The enhanced greenhouse effect

In Chapter 10 we explore issues related to global warming, associated with the enhanced greenhouse effect. There is no dispute that increased concentrations of the gases which act as a blanket around the earth, and which modify its energy balance, *can* have significant climatic effects. In this chapter we query, however, some of the evidence as to the likely *consequences* of the greenhouse effect, while in no way disputing the need for policies targeted on effective reductions in emissions which, in any case, may be environmental sensible. Our discussion is in terms of three dimensions to the greenhouse, emissions and global warming debate; they are:

(1) determining the facts—which *are* in dispute;

(2) measuring or predicting the likely *quantitative* effect of the enhanced greenhouse effect, and the alternative scenarios; and

(3) setting out appropriate policies and options.

Our conclusion, in brief, is that while there is a case for expecting global warming, there is not conclusive evidence that temperatures will rise significantly. The statistical models used are simply not robust, or capable of fine predictions. Without such evidence the prediction of *significant* global warming remains no more plausible than theories of the impending 'Ice Age' previously being predicted by some climatologists.

We note also that significant global warming could have both positive and negative effects, for example, increased concentrations of carbon dioxide can be beneficial through accelerating plant growth and lowering the water requirements for crops. If the greenhouse phenomenon proves well founded, both the positive and negative implications need to be assessed before concluding that the greenhouse effect poses a potential problem.

On the assumption that greenhouse gases are a problem, our preferred approach is one which favours reduced emissions through tradeable emissions quotas, possible carbon or emissions taxes, and other incentives to change behaviour. Policy decisions should be influenced by assessments of costs and benefits of different levels of increased greenhouse gas emissions under alternative policy assumptions. Furthermore, there is a general presumption that market mechanisms should be used in preference to 'command and control' strategies, since they are more effective devices than legislating the obligation to use particular technologies. If what we care about is emissions then we need incentives for firms to devise new and improved technologies, rather than impose known and no doubt dated and costly technologies.

Conclusion

The debate on environmental policy should focus on ways of bringing market based incentives to the fore in the environmental area. Our real priority, both for current and future generations, concerns the chronic failure of other economic policy settings to generate sustained economic progress. Government policies should, therefore, focus on removing inflation, getting interest rates to international levels, and generally promoting efficient use of resources through sensible taxation systems. In combination with the proper internalisation of external environmental costs, such an economic strategy has the capacity to make Australia healthy, wealthy, and environmental attractive.

2 A framework for policy

Executive summary

In the last twenty or thirty years a veritable arsenal of environmental policies have been proposed, discussed, and at times implemented. If this array of environmental weaponry is to be critically evaluated for its effectiveness, the objectives to which it is targeted first have to be specified. Ultimately, those objectives must focus upon efficient economic growth.

Efficient economic growth encapsulates the appropriate environmental policy goals. The notion of efficiency encompasses such matters as the 'quality of life' as well as the 'quantity of goods consumed'. It implies making the best use of resources, including natural resources and the environment, so as to maximise community well being. Efficiency requires carefully trading off the costs and benefits of resource use and environmental policy. It leads to policy recommendations which balance competing demands on resources rather than imposing total bans and moratoriums on productive activities.

Over the last two centuries, economists have given the concept of efficiency considerable intellectual rigour. In this chapter, insights from this body of thought are distilled and applied in an environmental context.

In an ideal world, efficient patterns of resource use could conceivably be determined by some omniscient planner. However, such planned utopias are impossible to achieve. Assembling and correctly weighing the requisite information is a dauntingly formidable task. The information itself is scarce and outcomes from its application are rarely certain. Moreover, bureaucratic managers have incentives to pursue political and institutional goals rather than maximise the productivity of the assets they control; furthermore, the constraints they face in the watchful use of valued inputs are more limited and therefore less compelling than those confronting private sector owners and managers. The risk that publicly owned or directed instruments will fail to protect the environment adequately is at least as real as the risks found with privately owned firms; the likelihood that publicly owned bodies will pursue the goal using excessive resources is much higher.

It follows that in a world of uncertainty, environmental policy should focus on providing broad boundaries within which the diverse knowledge and skills of individuals are channelled to seek out efficient environmental solutions. To achieve this, individual ownership and market-based incentives need to replace the political motivations of public resource owners.

Defining individual property rights to valued resources assigns liability in the case of anticipated and unforeseen circumstances. It creates incentives to adapt to changing economic and environmental circumstances. By defining boundaries rather than processes, individuals are given the flexibility to adjust and respond. In much the same way that genetic diversity and the processes of natural selection help ensure species survival in a shifting environment, so allowing individual creativity and entrepreneurship to seek out efficient market solutions ensures economic survival in a changing world.

The built-in incentives created by well defined, individual property rights have been recognised for centuries. Aristotle noted that:

> What is common to the greatest number gets the least amount of care. Men pay most attention to what is their own: they care less for what is common. (Politics, Book II, Chap 3.)

In many circumstances, fully defining and allocating property rights in environmental resources is sufficient to create the appropriate incentives for the owners of the resources to seek out their efficient uses. But vesting of property rights in individuals is not always possible, for example in the case of air where considerable definitional difficulties would be involved. Even if vesting of air were possible, this may well result in excessive transaction costs as the multiplicity of owners sought compensation and approvals from each other prior to embarking on any activity which might infringe on the rights to the resource.

A key aim of this work is the exploration of avenues where property rights approaches to environmental goods have a place. Where property rights solutions are not practicable, it does not follow that environmental problems should be redressed by regulatory means. The costs of regulatory approaches can outweigh any likely benefits. If regulatory approaches are to be considered, these should exploit market-based incentives as far as practically possible. Market-oriented policy tools allow efficient trade-offs and can save resources due to fully defined ownership. To this end, available approaches include:

- *taxes, fees and charges; and,*
- *setting limits to total emissions, whilst allowing resource users to trade their rights to pollute within these overall bounds.*

These mechanisms constitute a twilight between true market based approaches (in which the role of government as an independent actor is minimised) and 'command and control' approaches. Command and control approaches require a high degree of government intervention. They include direct government ownership of resources and regulations which define and limit technologies, processes and resource use. Their main deficiencies are two-fold. First, they offer inadequate incentives to the search for more cost-effective approaches. Secondly, command and control approaches suffer from a scarcity of information available to the bureaucratic planners who formulate the specific decisions; this severely hampers the abilities of such approaches to facilitate efficient resource allocation.

Efficiency of resource use

Markets and efficiency

Over the past two centuries economists have grappled with the notion of efficiency and sought to tease out the myriad implications of this far-reaching concept. Stripped down to its essentials, efficiency means making the best use of resources to ensure that community well-being is maximised. Resources include natural resources, capital, labour, knowledge and inherited institutions and cultural values. An essential element of the pursuit of efficiency involves taking energetic steps to reduce waste and to ensure that valued goods and services are provided with minimal cost. Equally important is the constant search for new and changed needs and consequential adjustments to outputs and inputs as these are discovered.

Meeting of needs requires trade-offs. We cannot have all the goods and services we want. Nor can we maximise the enjoyment of our leisure time without compromising our abilities to earn the where-with-all to afford it. What is to be maximised is human welfare registered by the decisions individuals make based on their own unique sets of preferences.

Accurately establishing the relative strengths of wants or 'needs' has proven possible only through the use of markets. Markets allow people to act according to their own judgements based on their preferences and their capacities to pay. The sum of these individual preferences for goods and services allows suppliers, conscious of their own capabilities, to determine supply responses capable of offering an adequate return. Competitive pressures also force suppliers to constantly re-examine their offerings and find less costly means of production and distribution.

An allocation of resources is said to be efficient if the resources cannot be transferred to other uses so as to make someone better off without, at the same time, making anyone else worse off. At the heart of this is a mutual willingness to trade on the part of the buyer and seller operating within an environment in which all things of value are individually owned. Only then can we be certain that both sides of the transactions are balancing the various opportunities they face, putting an appropriate effort into gathering and dispersing information, and adequately comprehending the risks and the opportunities foregone by making particular purchases and sales. Only when there is freedom to acquire and exchange commodities and information can we be certain that a mutual gain will emerge from transaction.

Economic efficiency is most usefully thought of as comprising two components: technical efficiency and allocative efficiency. Technical efficiency requires that inputs be arranged so that the output is produced at least cost. Allocative efficiency means the inputs are brought to bear in producing the pattern of goods which consumers value most.

Environmental attributes are part and parcel of the notion of economic efficiency. Transactions and resource allocations which do not take environmental concerns into account are unlikely to be efficient.

Goods and services which are to be consumed or owned *to the exclusion of others* have their worth verified by individuals' acts of purchasing; but this is a highly uncertain yardstick for those goods and services, such as parks and coastline, which offer wide benefits whether or not payment is made for them. Such jointly used goods, where there are difficulties in blocking access to those not contributing to their provision, are usually termed 'non-excludable' goods. The joint usage of many environmental services by a large number of people often makes it difficult to ensure that individual decisions embody the full consideration of environmental attributes. Where benefits of environmental amenities must inevitably be widely shared, disaggregated and individualistic market mechanisms may not always reveal their true worth. Because people will automatically obtain those benefits, they have strong incentives *not* to outlay their personal resources to obtain that access. In many cases, this calls for a more intrusive role for government than that of simply holding the ring in which transactions freely take place.

In making this point, it is important to bear in mind that environmental goods are not alone in having characteristics of non-excludability. Almost all transactions have spillover effects on non-contracting parties. One's neighbour's remodelled house will impact on the pleasure and perhaps the value of one's own house. An influx of new neighbours from Asia or southern Europe is likely to generate a supply of different restaurants which widen the diversity of choice for those previously living in the area. Widespread intervention to factor in all these spillovers would destroy the efficiencies and opportunities which markets create.

At issue is when government intrusion is necessary and how it is to be delivered. Discovery of true needs other than by market processes is very difficult; provision of these needs by government agencies will normally require excessive resource use because, unlike firms in competition with each other, government agencies do not face the same imperatives to constantly search for changes in needs and cheaper means of meeting them.

Values and efficient resource use

The valuations placed on uses of resources when determining whether allocations are likely to be efficient are ultimately those of individual men and women. While some individuals might obtain 'utility', or 'well-being', solely from the *quantity* of goods consumed, most, if not all, are interested in the *qualities* of goods, and almost everyone in developed market economies consumes many more services than goods. There is no foundation to the notion that economic efficiency focuses on the quantity of resources consumed, with an implied undervaluation of the 'emotional', 'aesthetic' or other less tangible aspects of human existence. Nothing in economics says that only 'material' or 'functional' uses of resources are to be valued in determining standards of living.

In fact, many of the goods and services supplied in markets are demanded for non-material reasons and a decreasing proportion of our expenditures are motivated by *basic* needs fulfilment. Our expenditures are not on food, warmth and shelter. They are on appealing to taste sensations as well as nutritional needs, on a surfeit of controlled heat, on well furnished houses with reserves of space. Most people do not go to restaurants because they are hungry, although the need to eat is one source of their demand for restaurant meals. Most of the huge annual expenditure on clothing, grooming and cosmetics is clearly motivated by non-material considerations. While some people demand education for the utilitarian motive of increasing their future earning capacities, many also are interested in, and pay for, education as a consumption good. Quantities of all these goods are simply means to ends—which are partly aesthetic.

Environmental goods and equity

If market freedoms are to be overridden, economic analysis can offer guidance on means of effecting this which are less likely to generate waste and distortion. Government reassignment of goods requires one party to relinquish some of his or her assets or income so that another can have more. In principle, decisions of this nature are based on judgements that some people's benefits over and above the resources they feel able to forego for them, outweigh the disbenefits accruing to those called upon to supply the resources. Political/bureaucratic judgements of this nature involve considerable hazard. Even where soundly based redistribution judgements are possible, these measures typically involve a sacrifice of efficiency. Redistributive measures also generate evasive measures on the party being taxed and encourage potential recipients to position themselves so that they obtain a share of the additional income in the offing. These sorts of behaviour will mean goods, services and energies are channelled away from their most productive uses.

It is often suggested that equity within and between generations should also be a goal of environmental policy. This concern with equity arises because a conflict is perceived between economic growth and the maintenance or improvement of environmental standards. As with all aspects of human life, there are many choices to be made between material goods and preservation, clean air, etc. Nonetheless, economic growth has been shown to be by far the most effective way of relieving poverty, cleansing the environment and providing social mobility between and within generations. Environmental goods tend to be more highly valued by individuals and societies with higher levels of income, and at the same time, higher levels of income make their provision more affordable. Krutilla and Fisher (1985) have cited econometric evidence (Boyer and Tolley, 1966) showing that environmental goods are 'luxury' goods, that is, demand for such goods increases more than proportionately with income. While Pearce found no evidence of 'willingness to pay' increasing with income in his 1980 survey (Pearce, 1980) he has shown, a possible relationship between what he refers to as 'willingness to act' and per capita GDP (Pearce et al., 1989). The relationship is especially clear when comparing developed and developing nations. Economic development may mean sacrificing some pristine natural endowments, but the alternative will mean less income and, inter alia, a diminished capacity to afford other environmental goods.

These considerations aside, it is unwise to focus on redistribution as an objective of environmental policy. Tax and welfare policies are more direct means of addressing distributional concerns. Environmental goods and services are, too narrow a component of overall needs and environmental policy is too remote from the vast complexity of redistributive goals to be a well targeted instrument for the achievement of satisfactory outcomes in this direction.

Additionally, the conflict between efficiency and equity is far less than commonly believed, since removal of the distortions and privileges which result from political intervention increases both efficiency *and* equity. In a free market economy, firms earning super profits or 'rents'—incomes above the minimum needed to ensure the continued supply of the goods they control—are always vulnerable to competitive pressures from others who could supply substitute goods or services. One way of preserving these rents is to lobby government to prevent entry to an industry or market. Thus, tariffs and subsidies, import quotas and licenses, industry regulation, government enforced monopolies, licensing of professionals and union monopolies all reduce the efficiency of resource allocation while at the same time creating contrived privileges.

In the environmental arena, government tolerated or mandated interventions are often inimical to equity. Thus, the provision of recreational facilities at low or zero prices, and disallowing those same resources from being employed for alternative uses, is a source of privilege to the mainly middle-class consumers of many of those environmental services. This source of privilege often directly disadvantages poorer individuals such as forest and mine workers. Its indirect effects may be felt more widely where they reduce the income of the community and adversely affect its capacity to provide resources to ameliorate the living conditions of the least fortunate.

The main objective

The focus of this work is anthropocentric, it is based upon *human* welfare. Although mankind arguably only has *custody* and not ownership of the earth, and although consideration may be due to all living organisms, the well-being which must dominate our conceptual framework is that of humanity and its progeny. Accordingly, we advocate the goal of efficient economic growth or, alternatively, efficient increases in the standard of living, as the appropriate objective for environmental policy. The question is: what mechanism will best ensure the meeting of that objective? It is here that markets, incentives, covenants, information and the efficient and fair vesting of property rights can play a crucial role.

Information, incentives and government intervention

Interconnectedness of markets and environmental goods

It is axiomatic both to ecologists and economists that everything is connected to everything else. The ramifications of a rise in demand for a good have been explored by many economists. Friedman and Friedman (1980) offer the example of an increased American demand for pencils changing global employment and price signals across a range of activities. The impact of an increased demand for pencils will extend to timber operations in Asia, graphite production, woodworking machinery and so on. At the same time it will lift costs of seemingly unrelated goods which make use of the same materials and thereby reallocate resources, perhaps making some redundant and bringing others into use. Like a sudden shout in a valley, the echoes will continue to have an effect long beyond the point where they are measurable.

A similar interconnectedness is to be found in nature. A change of ocean current may mean some micro-organisms reproduce themselves at a diminished rate, impacting upon the predator chain as other animals adapt or diminish in numbers. The loss of a piece of forest displaces and sometimes leads to the local eradication of whole species of wildlife, the genetic structures of which have become chained to the area and its attributes.

Government as an environmental regulator

The environment and the economy are complex interactive systems, involving essentially infinite volumes of information from individual actions. These informational complexities leave governments ill-equipped to form and implement detailed intervention strategies to improve people's quality of life. For government successfully to marshal environmental resources, at least two conditions must apply. *First*, they must have accurate information about how the environment will respond to possible changes in incentives and regulation and, in turn, about people's preferences and possible responses to changes in the economy and the environment. *Secondly*, policy makers must themselves have the appropriate incentives to consider all relevant costs and benefits and to ignore the mere jockeying of interest groups. The probability of governments getting it wrong is large, not least because of the high stakes to a few powerful players. Government failure is, we argue, far more endemic than market failure in the area of management, including pricing and allocating resources, precisely because of the informational and incentive difficulties inherent in non-market systems.

To apply economic analysis to resource allocation without the benefit of markets and prices, decision-makers must attach values to the margins of use. Such 'scientific' management requires that these values are known before an efficient solution can be calculated. Armed with this knowledge, the decision maker would seek to acquire the 'correct' information about resource values in alternative uses and to reallocate until marginal equalities hold. The planner's management problem is one of finding the socially optimal allocation. Schematically attractive though this process may appear to be, in practical application it is fatally flawed because it overlooks the vital importance of information. Both the vision, and the naivety, of the traditional resource economics perspective on information is captured by the economist Thomas Sowell (1987):

> Given that explicitly articulated knowledge is special and concentrated . . . the best conduct of social activities depends upon the special knowledge of the few being used to guide the actions of the many. . . . Along with this has often gone a vision of intellectuals as disinterested advisers. . . .

> If knowledge of values that must be traded off against one another were 'special and concentrated', scientific management

might be possible. But, as Nobel laureate F.A. Hayek has pointed out, this is not the economic problem which must be overcome in order to maximise the gains from human interaction. As Hayek puts it:

The economic problem of society is . . . not merely a problem of how to allocate 'given resources'. . . if 'given' is taken to mean given to a single mind which deliberately solves the problem set by these 'data.' It is rather a problem of how to secure the best use of resources known to any of the members of society, for ends whose relative importance only these individuals know. Or, to put it briefly, it is a problem of utilisation of knowledge not given to anyone in its totality.

The information and knowledge necessary for effective and 'scientific' management trade-offs is only revealed through human action. It is the very diversity, depth and breadth of differing kinds of information produced through exchanges between individuals which is lost through imposed and centralised actions. How can we know how much individuals truly value recreational opportunities, for example, other than by observing how much they are willing to pay for those activities? Merely asking them is unlikely to produce an accurate answer, given that they are not required to pay the cost for indulging their stated preference.

This can be graphically demonstrated by analysing the effects of studies using market research techniques (contingent valuation) to determine the value placed by the community on non-development of a resource. Careful testing of the proposal to mine at Coronation Hill (Resource Assessment Commission 1990) suggested the community would willingly offer a total of $650 million per annum if the site was left unmined. Doubtless other contingent valuation studies would throw up similar numbers to those found in the Coronation Hill study. For one square kilometre of country with no outstanding features, the Coronation Hill result could be construed as placing a value on non-development of Australia as a whole at something over 10000 times the nation's gross national product! There are scores of sites with attributes at least comparable to Coronation Hill. It is not plausible to imagine, given some 20% of Australia is already set aside as wilderness or aboriginal reserves, that the community would pay an additional $65 billion per annum to see 100 more small sites preserved or even $6.5 billion yearly to see 10 more such sites preserved. Certainly no political party would ever go to the electorate promising to tax and spend such incremental sums.

Even if non-market methods of determining choices can accurately reveal true community preferences, the political process of reports, submissions and inquiries which attempts to evaluate public response to alternative management plans is not costless. These informational and coordinating costs of the political process need to be subtracted from any possible gains from a more efficient publicly managed allocation of resources.

In practice, government committees have a tendency to base decisions on simplified, aggregate information. Rational individuals operating in this manner will frequently arrive at a ranking of choices which bears no relation to their individual preferences. This has been demonstrated mathematically in the so-called 'Arrow impossibility theorem' (under which a majority of rational people on a committee can be shown capable of preferring A to B, B to C, but then vote for C over A!).

The foregoing is not to suggest that governments are incapable of wise and rational decision making, but it is to suggest that they should do so within constraints as rigorous as those operating in a market. In addition, when governments do decide on an apparently wise course of action, it is often impossible to implement the policy in an efficient way, given perverse incentives, and the difficulties of motivating and monitoring bureaucrats and employees in government enterprises. For private firms, the bottom line—profit—provides a residual between the sums people will willingly expend on the products and services provided and the sums assembled to make them available. The measuring rod of profit—a relatively unambiguous criteria—is either absent or at best distorted in government enterprises.

Are there any gains from political resource allocation?
What are the conceivable sources of gains from political as opposed to private management of resources? If individuals can enjoy benefits from scarce resources without making a payment, market provision will not reflect the true benefits. Many con-

sumers may well 'free ride' on services provided for others and the provision will receive inadequate compensation for the services, because fewer of them will be produced than would be justified. For example, suppose many consumers enjoy having easy access to scenic views, but that once the access is available the benefits of the view cannot be restricted. Because individuals cannot be prevented from accessing the view, they have every incentive to understate their enjoyment if they are asked to contribute to its preservation or improvement; indeed they will look forward to a 'free ride'. Potential suppliers of such amenities (or those who would forego using them for other purposes) will find fewer willing customers than they would if those who did not contribute could somehow be excluded from access. There will be a tendency, then, to under-supply 'non-excludable' goods and services which are positively valued, and to over-supply those, such as many forms of atmospheric pollution, which are negatively valued. The market fails in these circumstances, to provide the correct information and supply incentives. However, as previously discussed, market mechanisms will rarely elicit true valuations when individuals cannot be excluded from the 'goods' or 'bads' which are produced.

Somewhat more problematic is the case where exclusion is possible, so that markets can work, but where there is no congestion from additional consumers. Suppose, for example, that consumers could be excluded from enjoying a view when they have not paid an 'entrance fee', but that the number of people wishing to enjoy the view is small enough for additional people to be allowed in without affecting the experience of existing customers. In that case, while exclusion is possible, it is inefficient on one score, because it would cost nothing to allow 'free-riders' in to enjoy the view. Nevertheless, if the government were to take over and provide free access to the view—which would mean access to all—the failure to charge would mean a loss of information about the value people place on that good vis-a-vis alternative uses of the same resources. It would mean that no solid information was revealed to indicate what additional sites of similar value should be preserved or created.

For reasons already addressed, efficient provision of the good by the government would also be unlikely. Even if management by the political process did capture an efficient allocation of resources in one instant of time, changing preferences, new technologies and different perceptions of proven resource availabilities will all necessitate further rounds of information gathering, thereby escalating the costs.

Inefficiency in planned economies

Information, and especially an asymmetry of information between supervisors and subordinates, is also central to explaining the appalling inefficiency of centrally planned economies. Managers of many government-owned enterprises in these economies know more about production technologies than do the planners, but many of them attempt to use that information advantage to do as little work as possible while paying themselves as much as possible. As Bernstam (1990) notes: In relation to the USSR,

> Like a bad secretary who receives less work from his/her boss than a good secretary, the regulated monopolistic firms are doing well by doing bad. They raise costs, fail new technology, waste resources, and otherwise maximise inputs in order to both justify the cost-based price increases and sabotage the state output pressure.

The waste and inefficiency involved in production in centrally planned economies has demonstrably been accompanied by environmental degradation on a scale far greater than that found in market economies. The fact that the goods and services produced are often not the goods and services which are most valued by consumers represents another major source of waste and inefficiency. This has its parallels within state owned enterprises in Australia. As argued by the Industry Commission (1990), many government-owned enterprises in Australia are very inefficient, and for reasons similar to those creating inefficiency in socialist economies. The inefficiency resulting from government ownership, regulation, and reduced competition, in many sectors of the Australian economy exacerbates environmental problems in that more resources are consumed than is necessary to provide our current standard of living. Given that the government sector at large accounted for 38% of national income in 1988–89, these wastages are a major contribution to any environmental damage associated with industrial activity.

Do governments look after future generations better ?

While it is true that any rational market investor will discount future relative to current costs and benefits, markets are in fact more likely to look after the interests of future generations than are governments. Politicians' time horizons can often be extremely short, and driven by headlines and public opinion polls. In Australia, there is a Federal or State election almost every year. Future generations do not take part in elections, but they are represented in the capital market. The future sums they are deemed prepared to outlay to enjoy future demands are likely to be quite diverse; this diversity reflects the great variation in appraisals of future worth likely to be held by a large number of individual owners of property.

In contrast to this diversity, government decisions are more likely to oscillate between one extreme or another. Democratically elected governments have a tendency to reflect the wishes of the marginal voter in the currently marginal electorate. Markets, on the other hand, can simultaneously reflect many of the more extreme views on the future value of a resource. Only markets can systematically cater to minority, and indeed minutely represented tastes.

Since the value of an asset hinges on expectations of what some others may pay for access in the future, in market economies those investing in future outcomes—often labelled speculators—become the representatives of future generations in today's markets. They are able to perform this function even when the prevailing view of their own generation is that such actions are not worthwhile. For example, a large number of now famous works of art have been preserved by the actions of private investors over many centuries, despite their contemporary unpopularity. Someone with an extreme, and strongly positive view about the great value of preserving a resource for future use, can bid for control of the resource in the market place, but might be ignored in the political system, or labelled an irrelevant extremist. As an example, private endowments decades ago preserved the nestings of eagles and other birds of prey when both agriculturalists and their contemporary environmentalists saw such actions as harmful. Present day environmentalists take a different view, and the diversity possible from such actions has been vindicated.

More generally, relying on individuals to provide for future generations does not mean future living standards will be ignored. An important component of the standard of living of the current generation is their view both of their own future consumption *and* the consumption of the next generation. If we leave intertemporal resource allocation to market processes, the evidence suggests that most people will in fact provide for an *increased* standard of living for the next generation, even though they discount future relative to current consumption.

It appears to be an almost universal human desire to provide a better life for succeeding generations. The enormous sacrifices of countless immigrants to Australia, the US, Canada, New Zealand and similar countries are testimony to the bequest motive. Other telling testimonies are the comments of many citizens of Poland and other Eastern European countries, interviewed on television over the last year, noting that the transition from central planning to a greater reliance on markets may involve short term hardships but will yield great dividends for future standards of living, particularly those of the next generation.

In short, direct government control of resource allocation decisions has proved most ineffective as a means of promoting increased standards of living. On the other hand, private ownership and markets give individuals a strong incentive to minimise the waste of scarce resources, to discover more highly valued uses of those resources under their control, and to preserve the value of assets.

Spillovers and risk

There are those who mistrust market determinations of the appropriate degree of the trade-off inherent in environmental protection. In part, these objections to allowing free rein to market forces stem from imperfections in market operations where goods are indivisible or non-excludable or where the vesting of ownership is difficult to envisage. In this book, we recognise that such claims and fears are not without substance; but we also recognise that government failure is likely to be as much, or more of a problem.

Some environmentalists would wish to see political action favouring preservation of some areas notwithstanding community wishes to the contrary. Often, because of the absence of markets and ownership, community wishes are difficult to determine and, as the previous discussion on the Coronation Hill proposal indicated, only those with a remote interest in something will find it difficult to envisage actually having to forego income for it and will therefore tend to overstate its value vis-a-vis the value they would place on it as a market good.

The spillovers or externalities resulting from mutual transactions are in the main adequately taken into consideration by market processes. There is certainly no alternative system available which can allocate goods and services more efficiently. And in spite of the exponential growth in transactions, as the human population has increased in wealth and numbers, the negative externalities have been held in check. In part, for example with air pollution, this has been due to government intervention limiting emissions of pollutants on behalf of the community. But overwhelmingly, measures to combat negative externalities have been made affordable by the increased incomes best arranged by free markets and limited government.

One element of these externalities is risk. For the generality of everyday living, risk accompanies our every actions. Cars create accidents on a scale unknown before the internal combustion engine; certain chemicals concentrated together in industrial processes threaten human life and the integrity of the surroundings which we value; new foods introduce exotic substances some of which appear to cause cancers. Yet life has become safer notwithstanding these changes. In fact, the pre-industrial woodfires we used for warmth, the 'natural' foods we ate prior to large scale cultivation and processing and the housing and workplaces of earlier millennia all posed much greater risk than those goods that have replaced them. Indeed, Ames (1990), one of the pioneers of testing for carcinogenic substances in foods, points out that 'natural' foods are just as likely to have carcinogenic properties as the synthetic substances over which considerable alarm is registered.

Notwithstanding the clear evidence that wealthier is safer and that the key to wealth is a narrow ambit for government actions, there are great concerns about the risks which market activities may pose. Wildavsky (1990) points to the near hysteria found in modern society about risk, he says:

The richest, longest lived, best protected, most resourceful civilisation, with the highest degree of insight into its own technology, is on the way to becoming the most frightened. (Wildavsky, 1990: 120)

In the forefront of concerns about risk are concerns about massive changes to the earth's ecosystem resulting from mankind's activities. Figures like Dr David Suzuki, operating through a media concerned more with sensation than objective analysis, have popularised grossly amplified versions of these risks.

There is assuredly a role for government to take decisions on behalf of the community as a whole and to undertake actions for which market mechanisms are unsuitable. National defence is a clear case of such a public good and measures to control urban pollution have already been mentioned as another. But the application of resources in this direction itself carries great risk in denying present and future generations higher living standards. When we divert resources from market activities, we reduce our abilities to afford to combat disagreeable facets of life in the future. Much is made of a putative inter-generational inequity in present generations possibly causing a change in the atmospheric conditions through modifying the ozone filter and greenhouse blanket which form the building blocks of existing living conditions. The jury is still out on the effect of mankind's activities on global temperature levels. But Wildavsky reverses the populist sentiments of doom promoters in arguing:

By what right, one might ask, does anyone enrich himself by enhancing future generations? By what right, it may be asked in return, does anyone impoverish the future by denying it choice? What would you have to believe about the future to sacrifice the present for it?...The future can acquire new strengths to compensate for old weaknesses, it can repair breakdowns, dampen oscillations, divert difficulties to productive paths...Fear of the future is manifested by pre-empting in the present what might otherwise be future choices. The future will not be allowed to make mistakes because the present will use up its excess resources in prevention. (p 123)

Risk accompanies our every action but government control and regulation is resource hungry and enterprise sapping. Before we allow these forces to conscript themselves into combating risks, we must be clear on the potential payoff.

Applying market solutions to environmental goods and services

The role of property rights

Trade and market operations require that property rights be defined and enforced. An important reason for the industrial revolution beginning in England was the security of tenure for private property provided by that country's well developed and independent legal system. Conversely, a significant deterrent to economic growth in many developing countries today is that uncertainty about property rights means that entrepreneurs can have little assurance that they will be able to reap the benefits from investments.

If you make an extra effort in countries with well-defined property rights, you reap benefits. Even those who are less able, or disadvantaged, gain in societies with well-defined property rights—they are able to benefit from the existence of a more prosperous society, and a society where individuals can create individualistic niches of business opportunity.

Defining property rights has applications capable of maintaining adequate levels of environmental services. Many environmental amenities and services are currently under-valued in market transactions, because they are scarce yet freely available. The best solution to this problem is the extension of markets wherever possible. Many ecological factors and values would be routinely incorporated into economic decisions if property rights to natural resources were to be properly defined and allocated.

Environmental values can be made part of the calculus guiding economic decisions made by private enterprise. Thus, most farmers, although radically transforming the ecology which preceded their activities, already can be said to pay attention to many valued ecological factors in their farm management decisions. Soil nutrients are conserved (and augmented) and measures are taken to prevent soil erosion. The upshot is, of course, about as far removed from any natural ecology as can be imagined, but the rustic ambience of grazing animals and of crops meets with most people's standards of a satisfactory environment.

Examples of somewhat manicured 'natural' environments produced solely from the stimulus of the profit motive are also to be found in privately owned US forests. In those parts of the US where forest recreational opportunities are not heavily subsidised by governments, firms like the International Paper Corporation employ ecologists to manage its forests to produce hunting, recreation opportunities, bird watching and other environmental services in addition to timber. Also in the US, non-profit organisations like the Nature Conservancy own tracts of forest, some of which are managed to produce timber and mineral output in addition to environmental services. Unfortunately, forest management which pursues multiple purposes like this cannot successfully do so where recreational uses are provided free by government. Such free provision 'crowds out' commercial provision, and by providing services without allowing for their alternative uses to be tested by market arbitration. Too many of the free goods are supplied and the community loses by trade-offs being disallowed between some of the free goods and some other goods. Moreover, the free goods are unlikely to be provided as cost-effectively by government as would occur where market disciplines and profit maximisation instil cost-reducing incentives.

Under the legal framework of defined and enforced property rights, so essential to markets, private parties are assigned liability for any effects their activities impose on others. Governments, while often a source of inefficiency when they over-rule mutually beneficial private trades, do have a role in helping the courts define the legal framework; they also have a role in enforcing the decisions of the courts. Governments operate best when they seek only indirectly to foster increasing living standards. It is preferable to leave most of the responsibility for commercial activity to private entrepreneurs acting within the rules laid down by government.

When the legal framework is inappropriate, or poorly defined, standards of living are likely to be compromised. Many 'environmental bads', such as pollution, are over-supplied because resources, such as clean air or clean water, are being used without compensation to alternative or potential users of those same resources, such as people wishing to breathe the air, drink the water, or bathe or fish in the streams or oceans. When the air or water is not owned by anyone it is available to those who get in first. Conversely, many 'environmental goods', such as the preservation of wildlife or genetic diversity, may be under-supplied because the lack of property rights to these resources limits the net return to entrepreneurs.

In some cases, markets for environmental goods are possible and desirable, yet have not been developed, because governments have been remiss. They have not helped the courts define and enforce property rights over the products or by-products at the heart of the problem. In other cases, difficulties of definition and measurement limit our ability to monitor and control access to resources, yet the 'spillover' costs or benefits are significant. In these latter cases there may be net benefits of direct government intervention in resource allocation decisions.

Employing market forces as regulatory instruments

Quasi property rights: tradeable emissions

Uncertain rights between polluter and pollutee means that air and water pollution are unlikely to be effectively managed by laissez-faire market mechanisms. While there can be 'money in muck', the costs of negotiating common law agreements between the parties to emissions or effluent are likely, on most occasions, to be prohibitive.

Excessive levels of pollution may best be combated by the quasi-market solution of defining and allocating tradeable pollution quotas; these are *alternatives* not *additions* to direct regulations. For direct regulation to be cost effective, governments would need a detailed knowledge of the production technologies and abatement costs of individual firms. In a competitive market it is unlikely that firms will either readily reveal such information or, until obligated by self interest, even seek it out for themselves. In contrast, charges and permits create incentives for individual firms to discover cost efficient means of reducing pollution by reducing output or changing technologies. In this way, Government activity is limited to setting overall targets and monitoring levels of pollution. Unlike direct regulation, governments do not need a detailed knowledge of individual firms' production processes, they simply focus on the defined emissions.

Having defined and enforced a set of effluent quotas, governments should allow the quotas to be bought and sold in an open market. Firms in different industries, or different firms in the same industry, often face radically different marginal costs of reducing pollution output. Those which can reduce pollution output at little cost will have an incentive to sell their effluent quotas, while those which find it very expensive to reduce pollution could purchase additional quotas. The outcome will be a cheaper means of meeting the goals. There will also be some dynamic gains as firms will also face a price for polluting, and they can decide on the appropriate trade-off between investing in pollution reduction and improving the efficiency of other aspects of their production process. Firms which succeed in eliminating pollution can profit relative to competitors who have difficulty meeting the standard. Sorting out just who produces what is then left to the market place, modified only by the governmental imposition of emissions ceilings and tradeable quotas.

There is no valid reason for restricting the purchase of effluent quotas to those who intend to pollute. Nationwide voluntary organisations, such as the Australian Conservation Foundation or the National Trust could purchase quotas and retire them. They could obtain funds from donations, by providing excludable goods such as magazines, naming rights to rivers, scenic reserves, koala enclosures and so on[1], or from the sale of limited mining or timber rights on their own land. Some corporations might also find that purchasing effluent quotas only to retire them from circulation could be a very effective means of corporate promotion. Once property rights have been defined and enforced and trade in them is facilitated, markets will give large numbers of people a strong incentive to be very imaginative in exploiting any potential gains from trade.

Somewhat paradoxically, limiting access to 'environmental services', such as opportunities to harvest or observe wildlife, or the provision of clean air or water, is the key to facilitating markets and thereby ensuring the continued 'sustainable' supply of these services. More explicitly, ensuring that environmental services are only provided when something else is given in exchange will guarantee that such services are taken into consideration in market transactions. When goods or services are given away without charge, they are over-used and abused.

For example, cattle and sheep have fared so well as species not only because they are used to provide goods and services in demand but also because they are allowed to be owned and traded. Conversely, it is mainly because it has been illegal to own and trade in elephants, rhinos and some other species in some African countries that numbers have been reduced in order to make way for animals, such as cattle, which can be owned and traded.[2]

While we readily acknowledge that many environmental amenities and services are currently under-valued in market transactions, the best solution to this problem is the extension of markets wherever possible. Where such an extension of markets is impossible, policy interventions should nevertheless harness the power of markets to accumulate and utilise relevant information. This can be accomplished by relying upon tradeable quotas, zoning and covenant restrictions and other similar devices which still allow private ownership and trade.

Taxes, fees and charges

Where ownership of rights to pollute cannot be vested,

mechanisms involving the use of charging allow some of the flexibility inherent in market operations to be tapped. Such mechanisms have the potential to achieve pollution reduction at a lower cost than direct regulation, as long as they replace the existing command and control measures. If they only become a further addition to a plethora of such regulations, outcomes could, in fact, be worse.

Taxes give decisions about the choice of technology to the private parties. These parties are likely to be both better informed about relevant choices and more highly motivated to select the most cost effective choice than government officials. Taxes and charges are less flexible than vesting of tradeable rights because they do not allow economies to be made by those able to make them most cheaply. With perfect information on firms' abatement costs, it is possible that charging mechanisms and tradeable rights would yield equivalent outcomes as illustrated in the Chart 2.1.

The chart assumes an independently determined level of output. Reductions in effluents and emissions are achieved by employing different technologies to produce this same output. The two intersecting curves represent the marginal costs of abatement (MCA) for two firms with differing reduction costs. MCAs tend to increase with each successive unit of effluent or emission reduction. Consequently the curves are drawn as exponentially increasing functions of the level of pollution reduction. The X axis shows these levels in units of concentration such as parts per million. However, for the high cost firm the axis is read from right to left, as opposed to the low cost firm which reads in the more conventional manner from left to right. The total length of the X axis is set equal to the required level of reduction—in this case 10 units of concentration. The Y axis measures abatement costs.

Constructed in this fashion, the chart shows the abatement costs incurred by both firms for every possible combination which achieves the desired total reductions. For example, if a uniform standard is set at 5, so that both firms have to reduce emissions by five units, then the abatement cost of the high cost firm will be *b*, and those for the low cost firm will be *a*. The efficient combination of reductions is *e*, where the marginal costs of abatement for each

Chart 2.1
Taxes and tradeable permits compared with a uniform standard

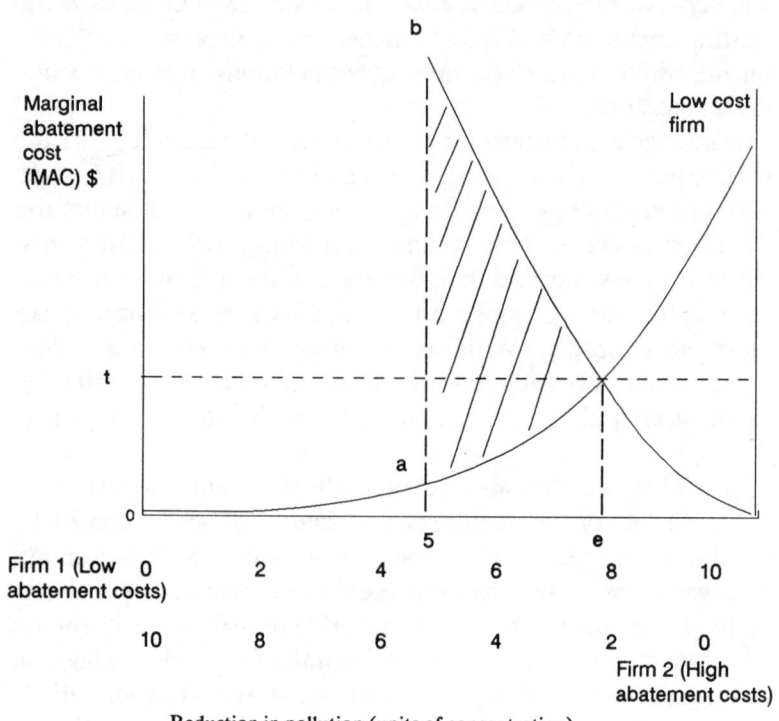

Source: Based on Tietenberg (1988)

firm are equal. At this point the total abatement costs are at a minimum. The low cost firm reduces emissions by eight units, while the high cost firm only reduces by two units. Compared to this cost effective solution the uniform standards imposes an extra cost on society equal to the shaded area. Regulatory standards could be varied but this would require knowledge of the individual firm's abatement costs which would be difficult and costly to obtain especially when the analysis is extended to more than two firms.

One alternative to attempting to individually calculate the

MAC curves of every firm, is to impose a tax on each unit of emission. After being first imposed at an arbitrary level, it would be gradually adjusted up or down until the desired level of reductions is achieved. Under the tax, governments would not require a knowledge of abatement costs but would need only to monitor emissions. As long as the tax is higher than marginal abatement costs, each firm has an incentive to reduce emissions. Once marginal abatement costs are equivalent to the tax paid on each unit, firms will choose to pay the tax rather than reduce emissions further. In Chart 3.1, t represents a tax of say 10 cents per unit of concentration. At this rate the high cost firm will reduce emissions by two units and the low cost firm by eight units. The tax achieves the desired level of abatement at the cost effective point e. Discovering the correct rate would require an interactive monitoring process but not the detailed knowledge necessitated by standards.

Tradeable permits represent an even more effective means of discovering the most cost efficient combination of reductions. Tradeable permits do not require an interactive process to search out the correct level of reductions. Instead, the total level is fixed from the outset at 10 units lower than current levels. Only permits up to this level are issued. The initial allocation could be the same as a uniform standard but firms would now be able to trade their pollution rights. Referring again to Chart 3.1 the high cost firm will be willing to buy his first additional permit for emitting one unit of concentration as long as the price is lower than his abatement cost 6. The low cost firm is willing to sell as long as the price he receives is higher than his abatement cost a. Both firms would benefit by trading at least one permit and by like reasoning they would also benefit by trading all their permits between s and e. If this trade is allowed to occur the efficient solution will once again be achieved. The advantage is that the level of reduction could be determined from the outset.

Other advantages associated with tradeable permits are that the level of reduction would remain constant over time as no additional permits are issued. In contrast, taxes would constantly need to be adjusted as population, and economic growth increased emission levels; and inflation devalues the tax.

The analysis so far has assumed that individual firms themselves know their marginal costs of abatement and the effectiveness of their abatement strategies. Baumol and Oates (1988) and others have pointed out that if there is uncertainty about marginal costs, and risks associated with either exceeding permit levels or over achieving reductions; then permits will not provide the most cost effective solutions. Nevertheless, uncertainty of outcome associated with taxes and the clear costs associated with uniform standards make a strong case for favouring tradeable permits.

Informational uncertainties aside, the conditions under which taxes and tradeable emissions could in theory be equivalent are very restricted. Taxes can only be equivalent to tradeable permits where the supply curve is fixed, a textbook fantasy in most industries.

When doing nothing is the preferred solution

In many cases both the costs of either defining and enforcing property rights and establishing the alternative regulatory processes, or the administrative costs of market mechanisms will turn out to exceed the net benefits of ensuring that all 'spillover' costs and benefits are accounted for. If the 'spillover' costs or benefits are not large in such situations, the best we can do is ignore them. This was explored some time ago in a seminal article by the economist Ronald Coase (1960). He notes:

> There is, of course, a further alternative, which is to do nothing about the problem at all. And given the costs involved in solving the problem by regulations issued by the governmental administrative machine will often be heavy... actions which give rise to harmful effects will be less than the costs involved in government regulation. (Coase, 1960: 18)

Similarly, when discussing the use of the common law to handle externalities, Coase notes:

> the reason why some activities are not the subject of contracts is exactly the same as the reason why some contracts are commonly unsatisfactory—it would cost too much to

put the matter right. (Coase, 1960: 39)

The key point of Coase's article is that there may be adverse consequences for efficiency if legal rights are assigned when transactions costs are not trivial.

In earlier sections, when addressing the vesting of transferable rights to pollute, it was argued that such a re-arrangement would lead to an increase in the value of production. But this assumed costless market transactions. Once the costs of carrying out market transactions are taken into account, it is clear that such an allocation of rights will only be worthwhile when the increase in the value of production consequent upon the allocation is greater than the costs which would be involved as a result of it being brought about.

In these conditions the initial delineation of legal rights does have an effect on the efficiency with which the economic system operates.

Coase argued that 'the ordinary law of nuisance' is likely to take these transactions costs into account and, that in order to maximise efficiency, we should assign the legal rights to one party in some circumstances and to the other party in other circumstances.

In yet other cases, the legal rights might be assigned to one party or the other by default, since an unambiguous assignment avoids the legal costs of deciding a particular case on its merits. For example, an individual might find his neighbour's garden to be incredibly ugly, but costs of determining the 'psychological harm' inflicted by a distasteful garden versus the 'psychological benefit' accruing to the owner of the garden from planting it how he likes, are likely to be quite large; a court would most likely decide that the costs exceed the net benefits from forcing owners of gardens to take their neighbours' preferences into account. From a legal perspective, such a judgement would rest upon the preponderance of the property rights involved. As a general rule, the owner of the garden would have the right to plant however he likes. If the decision were to favour the complainant, we would no doubt experience a rash of frivolous law suits as people acted in response to perceived deficiencies of their neighbour's gardens. The costs of litigation would, in general, exceed the benefits resulting from home owners taking account of their neighbours' preferences when managing their gardens. Ignoring the externalities will generally be the preferable approach to many such cases.

Transactions costs, or the costs of defining and enforcing property rights, include the costs of monitoring and measuring consumption or use, and the costs of negotiation and, perhaps, litigation. Monitoring costs are not frozen but will change, sometimes rising (where they rely on skilled labour inputs, for example) but more often falling as technological solutions are devised. Indeed, developments such as the silicon chip and video scanning are making it progressively easier to verify and measure actions. Moreover, private ownership of property creates incentives to devise privately profitable new ways of monitoring access and use of that property. For example, privately owned cattle herds grazing on vast acres of privately owned or controlled land created a market opportunity for the invention of barbed wire. This greatly reduced the costs of monitoring cattle, enabling increased scope for improving grazing land as other 'free riding' farmers and wild animals could be excluded from the benefits. Similarly, the electronic monitoring of vehicles which choose fast lanes, and the scope for computerised monitoring of car emissions across freeway entrances or other locations, have the possibility of internalising many costs drivers and cars impose on others.[3]

Negotiation costs are a further element of transaction costs. These include the time and effort involved in reaching mutually acceptable agreements and can involve the costs of lawyers, accountants and specialist brokers. The extent of formality in the negotiations, and therefore the costs involved, will depend on the number of parties to the transaction, and may be quite sensitive to how well the parties know each other. The costs will also depend on the legal precedents which have been established. If past judgements are clearly applicable to a current case, it is unlikely that the dispute would proceed to litigation. All parties could be confident of the decision the courts would reach were the dispute to be litigated and they would have a large incentive to avoid unnecessary legal costs.

Low benefits from defining property rights can also imply that the situation is best left alone. But again this situation is not immutable. Some sources of pollution, for example, may have been of less concern in the past partly because most environments are able to cope with small amounts of pollutants without serious consequences. This would suggest that it has become efficient to control pollution only as the amount of pollution has risen. Similarly, the demands for clean air and water, and recreation opportunities, probably have increased along with increased standards of living, and this has also raised the benefits of defining property rights to many environmental amenities.

Endnotes

1. Many of these natural and man-made sites are now named after politicians, the royalty or historical figures. They are marketable services, however, which could be used to raise funds, much as private universities in the US raise funds by selling naming rights to buildings and so forth.

2. In some cases, even if it were legal to own and trade in wildlife, the value of the wildlife might be less than the value of cattle so numbers might still be reduced to make way for additional cattle. We would expect, however, that as the numbers of wildlife species fell their relative value would rise. Many of the wildlife species probably also would be less expensive to farm since they are most likely better adapted to the local ecosystems. Thus, they may be chosen in preference to cattle even if their market value is somewhat less.

3. These technologies are not just pipe-dreams. Electronic toll and traffic management systems were implemented on a New Orleans toll bridge in June 1989, the Alesind Tunnel in Norway in 1987, and a number of other Norwegian, French and Italian tollways. The systems involve a small electronic card attached to your car which can be detected by a laser monitor. Extending the scope of the systems to include conjection pricing and pollution control are technologically viable possibilites. Robert E. Poole (1990) *Electronic Toll Collection — Key to Solving Urban Freeway Congestion*, Santa Monica, CA, Reason Foundation.

3 Managing the environment: an historical perspective

Executive summary

Current resource policy is strongly conditioned by our perception of the past. It is therefore important to understand the natural and institutional constraints under which the development of Australia's resources has taken place. Previous generations pointed to the triumph of converting a brown, parched and impoverished land into a nation supporting millions of people. Pride was expressed at the feat of creating a nation producing and marketing twice as much food and raw materials as it required for its own needs. In the process, a standard of living was forged on a par with the world's leaders. To many, self congratulation over 200 years of achievement has turned to remorse about the costs involved. Visions of Australia's European settlers clearing the land, cutting down the forests, and introducing exotic flora and fauna, and disturbing the native ecology, have caused many to suffer feelings of guilt. It is folly to be smug about the past. But these feelings of guilt are surely misplaced. Mankind has benefited from the taming of the Australian wilderness.

In the early 1800s, natural endowments were abundant relative to human and physical capital. Living standards were low. Often, in the very struggle for survival, the intrinsic values of the environment had a lower priority. By using Australia's abundance of land, mineral and timber resources, the graziers, gold miners and timber getters of the past were able to generate wealth to pass on to their grandchildren. Without this accumulation of physical capital and human knowledge it is doubtful that we would now be able to afford improvements in environmental quality or have the skills effectively to manage Australia's unique natural environment. In applying critical tests to the actions of previous generations, it is unwise to assume on their part the preferences we have today.

This chapter documents how the early development of Australia's abundant natural resources was consistent with the endowments, knowledge and market opportunities of the time. Environmental damage certainly occurred. Leaving the continent untouched by western civilisation meant leaving it in a form which could support no more than several hundred thousand souls living in poverty. Furthermore, in many cases any genuine damage resulted from an inadequate definition, or monitoring, of property rights; and from imperfect information, rather than from the process of development. The rapid depletion of early coastal fisheries is a prime example of inappropriate development caused by absence of property titles and open access. Part of the damage also resulted from ignorance of the nature of a new and unfamiliar environment. However, this ignorance was reflected as much in the governmental polices of the time as it was in the individual settlers themselves.

The chapter shows that certain government policies distorted market signals by subsidising infrastructure, biasing land use and contriving particular forms of settlement. These polices contributed to and exacerbated environmental damage. For example, closer settlement has been a dominant political goal in Australia since the mid nineteenth century. A large number of closer settlement and irrigation schemes were implemented without sufficient understanding or regard for environmental considerations. Soil degradation and salinity are two of the legacies of inappropriate land use which these schemes brought in their wake.

Environmental management over the past century has remained very much the province of governments in Australia, and regulation has been the favoured instrument of management, but the results of regulation have often been devastating. There are also an abundance of examples of ecologically sound private developments under properly designated property rights and covenants. But examination of the most effective mechanisms for allocating priorities between environmental preservation and productive activity has been neglected. This chapter addresses this neglect.

Introduction

Australia's historical record provides a vivid backdrop to our topic. Against this backdrop we can consider some of the forces and influences affecting patterns of resource use and environmental management. The firestick farming of the Aboriginals; the development of the early fishing, timber and pastoral industries; and the mining boom of the 1840s; are but some examples from a rich historical reservoir.

Mankind's transforming of the environment began, not with European settlement, but with Aboriginal communities. However, beginning with the first whalers and sealers, European settlement delivered a series of more rapid changes and developments in the Australian landscape and its ecosystems. Many have interpreted the historical record of European settlement as an indictment of 'free' markets and 'un-restrained' resource use. 'Resource exploitation' is an invidious charge levied against the colonial settlers. This charge overlooks four important factors:

- the future opportunities created by earlier resource use;
- the importance of trade-offs between resource use and preservation;
- weaknesses in the specification of property rights to resources;
- inevitable information deficiencies.

In the rapidly growing international economy of the nineteenth century, Australia's settlers behaved similarly to other peoples. Resources were used in ways consistent with their relative scarcities. In the classical taxonomy of factors of production, Australia was well endowed with land but short of labour and capital. In the manner observed by Adam Smith with reference to England's North American colonies, the Australian colonies achieved prosperity by combining investment and the skills and energies of immigrants from the British Isles with the abundant resources provided by the natural environment. The increasing demands for raw materials and food generated by the industrialising countries complemented the increased supply potential the new continent could provide (Smith, 1958: 63).

Resource use rather than conservation provided the foundation for Australia's future prosperity. Increased prosperity generated increases in the capital stock of physical assets, infrastructure, and human knowledge. These in turn enabled resources to be used more effectively, and generated new opportunities and developments. Paradoxically, using resources ultimately meant consideration of resource conservation could be afforded. After a certain level of development has been achieved, and particularly when a society is substantially urbanised, the demand for the preservation of the natural environment tends to increase as the satisfaction of more immediate needs of food and shelter have been met.

As well as allowing more environmental services to be afforded, much development calls for a trade-off between preservation and use. In this respect, abundant resources are not valued as highly for their preservation properties as scarce ones. Resources like forests, which now have scarcity value, were regarded as obstacles to development when trees covered land required for farming. When the income stream from farming exceeded the preservation value of trees it was rational to dispose of trees, just as it may now be rational to conserve them. How much disturbance to the existing environment a society is prepared to tolerate involves trade-offs between different land use options, including preservation of the native environment. The cost of environmental preservation is the stream of income foregone by not choosing developmental options. Most people want to have both development and a pleasant environment, but it is probably true that in the early stages of settlement people have a greater preference for development and tend to see the natural environment as needing to be tamed to meet their immediate needs. After all, they have plenty of natural environment and many needs to be met.

The tendentious charge of 'exploitation' is an accurate description of some early resource based activities like: destruction of valuable stands of huon pine and red cedar, and the decimation of whale and seal populations. In these instances, however, excessive use resulted from inadequately defined or monitored property rights. Patterns of resource use must be considered in the context of who owns and monitors the resource. The definition and enforcement of property rights is a crucial element in environmental management. Commonly owned resources, such as fishing resources, will be over-used when there is open access to all. Rights to resources which are assigned, but poorly monitored, will also be poorly managed, as in the case of native forests.

From the beginning of European settlement, rights to most resources in Australia were vested in the Crown. By their decisions regarding the alienation of land and access to resources the Colonial governments played a major role in determining the course of resource use. These interventions often inadvertently encouraged less than optimal economic use of resources and were instrumental in bringing about environmental damage.

Australia's environment was alien to it's early European occupants. To many it was a harsh taskmaster. Few had the information necessary to optimally manage its various moods and vagaries. For example, the East Strezelecki ranges in South Gippsland were known as the 'heartbreak hills' as settlers tried to come to terms with an environment unsuitable for farming. Mistakes were made; but, with no prior knowledge, it was doubtful these could have been avoided. Not only individuals and markets failed—a poor knowledge of Australia's ecology brought mistakes on the part of bureaucracies and governments too. Subsidised irrigation schemes have been a major contributor to problems of wetland salinity and the policy of closer settlement encouraged farming densities in excess of carrying capacities.

Unlike individual owners, governments do not have a personal stake in outcomes. Moreover, when the process of discovering the strengths and limitations of the land is independently conducted by individual farmers, the damage is likely to be far less extensive than if it follows some master plan. Experimentation leads to small mistakes and emulation of successful approaches. Major wholesale planning approaches contain the seed of potential large scale catastrophes.

Future opportunities, trade-offs, the specification of property rights and the importance of information, are major themes of this chapter. In a brief twenty pages it is obviously impossible to cover the depth and breadth of Australian economic history. Instead, we examine, in turn, the early history of major resource based industries and their environmental impacts. Where possible, some attempt is made to follow the chronological development of patterns of resource use. We start with the Aboriginal subsistence style land management, and the early whalers and sealers. We then trace the development of the early pastoral industry, the gold mining years and the move towards encouraging agriculture. Finally we consider the problems created by a rapidly growing urban population.

The early beginnings

The history of Australian resource management begins with the Aborigines. European settlers did not come to a natural environment untouched by human hands. For centuries the hunter-gatherer, Aboriginal societies had significantly modified Australian ecosystems through practices such as fire-stick farming and hunting.

The Aboriginals used the firestick as an essential aid to hunting, setting fire to scrub and grassland to flush out prey. The firestick was also a farming implement. Burnt-over land produced sweet, green shoots of grass and modified the vegetation in favour of grassland at the expense of scrub. It is unlikely that many parts of the Australian continent escaped regular burnings as part of the Aboriginals firestick husbandry. The grasslands and open woodlands, which were so attractive to the early European graziers, had been cleared of their scrub and forest cover as the result of the long period of environmental management by fire practised by the Aboriginals (Blainey, 1975: Chap 5).

Extensive modification of the environment by the Aboriginals supported little more than a primitive subsistence economy. The discovery of the great southern continent by the Europeans, in the late seventeenth century, heralded the beginning of a new pattern of resource use. Whereas the Aboriginal societies derived little more than subsistence, European efforts were directed towards producing surpluses to be sold in international markets and installing urban transport, and other infrastructure which also resulted in significant modification of the environment but produced a high standard of living for a much larger population.

Whaling and sealing

The earliest Australian exports came from the fisheries off the south-eastern coasts of the continent. The British government fostered whaling because it provided a good schooling for seamen and encouraged the building of strong ships suitable for long voyages. The convict fleet which arrived in Sydney in 1791 included five whalers which returned with cargoes of whale oil. Very few colonial investors had the capital to engage in the hunting of the deep sea sperm whales so most of the colonial whaling was confined to the black whalebone, or bay whale. These did not require ocean going ships, and which was fished from the Derwent estuary and Twofold Bay south of Sydney during the winter months when cows came into the sheltered waters to calve (Blainey, 1966: Chap 5). In 1809 Britain imposed a high protective duty against colonial oil, and until the early twenties British whalers dominated the trade until duties were removed. The great era of colonial whaling flourished for nearly two decades until the depression of the 1840s (Bach, 1982: 75).

Until 1835 whaling provided one of the most valuable exports of Australian produce. Sealskins were another product of the fisheries which were taken on the islands of Bass Strait and Kangaroo Island off the South Australian coast. As they came ashore to breed, the seals were clubbed and their skins taken. The slaughter of seals during their breeding season brought the rapid depletion of the resource and by the 1820s sealing was in decline. Although fisheries contributed significantly to the colonial economy, property rights were not defined and no quotas were set. Absence of property rights and other market mechanisms meant an absence of incentives to ensure the conservation of the resource. By the middle of the century commercial sealing grounds had been seriously depleted.

Timber from the forests

A similar lack of title and the absence of incentives to harvest rather than deplete were found in early timber gathering activities. One of the motivations in the choice of New South Wales as a convict settlement was the hope that Norfolk Island pine would prove to be a plentiful source of naval timber for Britain. Although this hope was not realised, timber was nevertheless essential for construction and maintenance in the settlement. It was also required for the repair of visiting ships and for construction in local boatyards. The authorities were on the look-out for useful stands of timber and explorers and settlers were encouraged to note and report on the character and availability of local timber supplies.

In 1795 Governor Hunter ordered that the King's mark be put on all trees which might have a use for building or naval purposes. He forbade the felling of timber on Crown Land which had not been marked out or allocated to individuals. The regulation also prohibited the cutting of trees, useful for naval purposes, on private land. In spite of heavy penalties for breaches of these orders, it seems that they had very little effect in curbing the illegal taking of timber (Williams, 1988: 124). Rights to the resource were defined but they could not be enforced.

The red cedar of the Hunter Valley and the coastal rainforest of New South Wales were highly valued, and the authorities seemed totally incapable of controlling cedar felling in the country areas. Gangs of cedar-cutters continually roamed along the coastal rivers looking for good stands and they formed the first settlements on the Richmond River in the north and at Kiama in the south (Bach, 1982: 121–2, 231). Exploitation by these roaming fellers was so thorough that the cedar stands of New South Wales were all but cut-out by the end of the 1860s. When the cedar was cut-out the timber fellers attention turned to the stands of hoop pine along the Tweed, Richmond and southern Queensland rivers (Williams, 1988: 124).

The other colonies fared little better. The highly regarded Huon Pine from around Macquarie Harbour on the remote south west coast of Tasmania was virtually cut-out by 1870 while in South Australia the stringybark forests in the Mount Lofty ranges were illegally felled. Timber-getters followed the land surveyors cutting out the best trees so that a selector, when he took up his land, might find it denuded of all valuable timber (Williams, 1988: 124).

In the beginning forests seemed to be an inexhaustible source of timber. The government was concerned to ensure supplies for their own needs and to raise revenue but, in spite of attempts at regulating the taking of timber, it lacked the resources to enforce the regulations, especially in the more remote areas. Farmers, on the other hand, saw forests as an impediment to their operations and the clearing of forests, in a community which earned its living from farming, was socially beneficial (Carron, 1985: 3).

As valued timber supplies dwindled, a greater premium was placed on conservation. This behaviour stemmed not from ecological motives but from perceptions of possible timber shortages. It lead the Colonial governments to pursue a policy of public management as well as Crown ownership of the resource. Many of the early Land Acts allowed for the setting aside of land for the preservation and growth of timber and in the last decades of the nineteenth century state ownership and management of forests became the prevalent pattern. For instance, in Victoria in 1873 over 600,000 acres had been set aside as state forests and by 1903 this had increased to almost five million acres (Dingle, 1984: 139). During these years every colony moved to some form of forest conservation by reserving state forests and setting up forestry departments to manage the resource. South Australia and New South Wales began reafforestation with exotic softwoods (Williams, 1988: 125).

The early years of the twentieth century saw the professionalisation of forestry in Australia. There were seven interstate conferences on forestry held in Australia between 1911 and 1924. A number of themes emerged from these conferences including a demand for advanced technical training, afforestation, reservation, significant forest cover in highland areas to promote reliable water conservation and erosion control, and fire monitoring and prevention measures (Powell, 1988: 160–8). However, little attention was paid to managing timber economically. Management practices borrowed heavily from European traditions which took no account of price or interest costs. Despite efforts at conservation practices, excessive emphasis on timber management for sustained yield contributed to the large operating deficits of the forests services and may have exacerbated later confrontations with environmentalists.

Opening up the interior

The first settlements in New South Wales were established as gaols. As a prime administrative consideration was the security of the prison, the early governors were anxious not to allow the settlements to extend beyond the bounds of effective control. On the other hand, because they were so isolated, another important objective was to ensure that the settlements were as self-sufficient as possible in foodstuffs. Accordingly, free settlers and ex-convicts were given grants of land and the assistance of convict labour.

Land for farming and grazing was the most immediately useful natural resource and for the first three decades settlement was virtually confined to the Cumberland Plain, bounded on the east by the Pacific Ocean and to the west by the barrier of the Blue Mountains. When the mountain barrier was penetrated in 1813 it was neither scientific curiosity nor shortage of land which provided the incentive, but a temporary shortage of pasture due to drought and a caterpillar plague (Perry, 1963: 27).

The crossing of the Blue Mountains revealed the existence of extensive pastures and, in times of pasture shortage on the Cumberland Plain, Governor Macquarie allowed stock to be taken over the mountains for the duration of the crisis to obtain temporary relief. The area made available for grazing was strictly defined and stock owners were required to vacate the area at one month's notice. Permission for a more extensive temporary occupation of the 'New Country' was granted by an 1820 order which recognised the growing pressure on the pastures of the Cumberland Plain and permitted pastoralists to take up runs in the south-western area of the Goulburn and Breadalbane Plains and around Bathurst.

In 1824 Governor Brisbane provided some security against trespass for these occupiers by issuing a 'ticket of occupation' which provided exclusive grazing rights to a certain area but restricted the right to cut timber to what was required for the building of huts and stockyards (Perry, 1963: 34). In addition to providing some security from trespass, without actually alienating the land, tickets of occupation enabled immediate occupation without prior survey and did not pre-empt the future disposal of the land.

A shortage of surveyors meant that the colony was only crudely mapped in the 1820s. Nonetheless, the Colonial Office in London, anxious to see more control exercised over occupation, issued orders for the colony to be divided into counties, hundreds and parishes (in 1825). These orders also required the land to be valued preparatory for selling and the Colonial government therefore ceased issuing tickets of occupation and sought to specify areas which could be surveyed and valued quickly as limits of settlement. The limits were announced in 1826 and revised in 1829 when the nineteen counties were proclaimed (Perry, 1966: 45).In proclaiming the Limits to Settlement 'the Government erected a huge non-trespass sign over the interior', to use the words of Stephen Roberts (Roberts, 1964: 4). It was a sign which was to be flagrantly ignored by pastoralists who streamed out beyond the Limits of Settlement and, stimulated by the growing demand for wool, 'squatted' on runs which soon dotted the map from the plateau of New England in the north to the lush pastures of Australia Felix in the south and extended as far west as the banks of the Darling River.

Seemingly powerless to prevent squatting, Governor Bourke set about legitimatising grazing outside the Limits of Settlement. He recognised the reality that:

> Not all the armies of England—not a hundred thousand soldiers scattered through the bush—could drive back our herds within the Nineteen Counties.

Nevertheless, he had difficulty in convincing his superiors in London (Roberts, 1968: 188).

Eventually Bourke prevailed and in 1836 the first Squatting Act was passed by the Legislative Council. This Act allowed stockmen to occupy Crown lands for the yearly payment of a ten pounds licence fee. This was amended in 1839 so that the squatter had to pay both a fixed annual licence fee and a tax on his stock. The proceeds were used to provide resources for the Land Commissioners to police the outer areas (Roberts, 1968: 189).

While in New South Wales, squatters were being enabled to occupy thousands of acres of land for a mere ten pounds (Roberts, 1964: 81–82), across the border, in South Australia, land was offered at a uniform price of one per acre. The colony of South Australia was founded on the Wakefieldian principles of systematic colonisation and concentration of settlement. The key to Wakefield's scheme was the 'sufficient price' for land. Wakefield argued that land needed to be priced at a level which was high enough to prevent farmers from purchasing more land than they could profitably farm and to prevent immigrants from buying land in their first seven years so that these would instead provide a pool of labour. Immigration was to be funded from the proceeds of the land sales (Roberts, 1968: 83–89).

An inherent conflict developed between the two colonies over land policies (Pike, 1967: Chap. 3). In London, where the Wakefieldians had the ear of the Colonial Secretary, pressure was applied to force New South Wales to adopt policies more in line with Wakefieldian principles. This campaign finally lead to the proclamation of Earl Grey's *Waste Lands Order-in-Council* in October 1847. Grey imposed a uniform upset price for Crown land throughout the Colonies. He was opposed to alienating land below this price and also to annual tenure. The *Order-in-Council* introduced long leases for pastoralists with the right to purchase at the upset price at any time. One of the effect of this new policy was to allow true ownership by squatter farmers and to bring genuine incentives for long term custody.

Pastoralism

The squatting system proved to be an appropriate arrangement for the development of the pastoral industry which by mid-century was by far the greatest source of Colonial exports.

The early pastoral industry was capital intensive, but required extensive tracts of land at nominal prices. The squatter's capital was invested in livestock and stores. It was only after the gold rushes and increased competition between pastoralists and settlers for the better lands in the south-east of the continent, that the pastoralists were forced, by a shortage of labour, to invest in substantial station improvements which raised the productivity of their more highly valued lands (Heathcote, 1965: 88–91).

Squatting on unoccupied Crown Lands meant bringing into

production land with an opportunity cost of virtually zero. Many regard the squatting system as being a very flexible form of tenure because it enabled the land to be productively used while leaving options open for alternative, more intensive later use. At a time when the native environment was not well understood, the extensive grazing was less disruptive than more intensive land use, although the introduction of exotic animals was bound to cause dramatic changes in the native vegetation (Moore, 1962).

However, the absence of permanent tenure contributed to overgrazing and the associated problem of drought. Australia is a drought prone continent and this must be a prominent consideration in any environmental management scheme. Too often, in plans for closer settlement, drought has been regarded as exceptional. The fact is that drought is a regular and almost predictable occurrence.

It took years, often decades, of experience and observation for pastoralists and settlers to become acquainted with the impact of drought and flood on the carrying capacity of the land. Settlers often moved into an area on the basis of favourable explorers' reports only to learn the harsh reality that the good seasons are followed by the bad ones (Perry, 1966). Where land was not permanently vested, there were inadequate incentives to exercise caution in stock levels.

Moreover, bureaucrats in Departments of Lands offices drew lines on maps and planned closer settlement schemes which were often quite inappropriate for the local climatic conditions. The Warrego country straddling the central border between New South Wales and Queensland was a case in point. From the late 1840s this region had been occupied by graziers under pastoral leases. In 1884 the Queensland and New South Wales governments passed legislation aimed at the subdivision of the larger pastoral properties in order to promote more intensive occupation and development of the plains.

This attempt to permanently settle 'an industrious population on the public estate' failed because of unrealistic assessments of the carrying capacity of the area. The result was a deterioration in the native vegetation which reduced the carrying capacity of the Warrego country. The long drought of the 1890s took its toll on the native vegetation. The smaller graziers found their runs were overstocked. Perversely the provision of watering facilities added to the overstocking and rabbits exacerbated the problem. By 1901 it became clear to pastoralists that the stocking rate during drought was not limited by the supply of available water, but by the availability of feed. The small grazier who lost his stock during drought usually had insufficient capital to replace it when the drought broke (Heathcote, 1965: Chap 6).

Graziers, no less than bureaucrats were capable of over-estimating the carrying capacities of native pastures. The practice of 'flogging' runs bare was sometimes followed, no doubt on a wider scale than would otherwise have prevailed if tenure arrangements had been more secure. Cases of 'flogging' a neighbour's run were not unknown to the courts (Williams, 1962).

The seven year drought which afflicted the Western Division of New South Wales from 1895 to 1902 not only forced home the realisation that long droughts were characteristic of inland Australia but also indicated that pastoralism was not a suitable occupation for small graziers lacking access to large financial resources. The survivors of this long drought were the larger farmers and pastoral companies which had the flexibility and financial resources to move stock between properties located over a wide area and to restock the properties when the drought broke. By avoiding gross over-stocking during drought the larger enterprise was able to manage the relationship between the pastoral activities and the arid environment more effectively so that edible vegetation was not replaced by inedible varieties (Cain, 1962).

Although pastoral lessees on the rich plains of the Riverina in New South Wales and the Western District in Victoria were able to convert substantial portions of their leases into freehold title, the major part of the land in the more arid parts remained unalienated. As late as the mid–1960s, only a little more than 10% of the Australian land area had been alienated or was in process of alienation (Campbell, 1970: 172–173). The pastoral occupation of large areas of Queensland, the Northern Territory and Western Australia is under leases which generally run from 25 to 40 years.

Retaining the land under Crown ownership keeps open the option of reassessing property sizes in the light of economic and technological changes. It does, however, remove much of the incentive for lessees to develop the properties.

The South Australian government had been keen to develop the pastoral potential of the Northern Territory. To this end it introduced a leasing system in 1890 which provided generous terms for the lessee. In relation to stocking, only one head of cattle per square mile was required within three years of the application for the lease and two heads within seven years. Rents were set at sixpence per square mile for the first seven years of a lease, one shilling for the second seven years while for the remainder of the 45 year lease the rent was to be fixed by valuation.

These low rents however were insufficient to generate the financial resources necessary for the proper servicing of the runs with communications, water boring, conservation and policing. Furthermore, the same rents applied over the whole area, failing to differentiate between land of lower and higher stock-carrying capacity or between varying distances from markets. The provisions relating to tenure and compensation for improvements were also unsatisfactory. The South Australian government was unable to effectively address the problems of management in an area where pastoral enterprise was extremely hazardous and effectively placed obstacles in the way of private enterprise attempting the task (Duncan, 1967: 116–122).

Controlling exotic vermin

While the pastoralists were still discovering the optimum levels of stocking intensity on the native pastures of the semi-arid interior, they had to face the devastation caused by advance of rabbits across New South Wales and Queensland during the 1880s. The effect of the rabbit infestation was to severely reduce the stock carrying capacity of the pastures. In the Western Division of New South Wales carrying capacity was reduced by about 50% between 1891 and 1911. The combination of maximum stocking intensity and the rabbit advance inflicted irreparable damage on the native pastures which left the country very susceptible to erosion (Fennessy, 1962: 228). Although governments had the power to adjust the terms of leases to arrest deterioration in vegetation or to prevent soil erosion, there is little evidence that the relevant departments had sufficient knowledge of the management of native vegetation to institute rational controls over grazing (Campbell, 1970: 182–183). A weakness of the leasehold system became evident in the Riverina in the 1880s when squatters faced the rabbit invasion from Victoria. Without secure tenure of the holdings and with no property rights in the improvements they constructed, squatters were in no position to invest in, say 100 miles of rabbit proof fencing at 50 per mile, when their lease could be withdrawn at short notice without compensation (Rolls, 1977: 146). Fencing would have reduced the impact of rabbits but, because of monitoring problems, there were limits to its effectiveness.

All governments in Australia have relied heavily on legislation to deal with the problem of vermin control. Following the recommendation of a 1889 Royal Commission in New South Wales, the principle of the obligation of the individual occupier for the destruction of noxious animals and plants became firmly established in Australia (Sawyer 1962: 243) The principle of 'landowner onus' may have been firmly established, but that was no guarantee of effective enforcement. Often the enforcement agencies, like the Pastures Protection Boards which have had responsibility since 1902 for controlling rabbits in New South Wales, were so ignorant of the life history of rabbits as to be ineffective and their efforts achieved little more than the needless poisoning of native wildlife.

While governments passed legislation to force landholders to eradicate rabbits, unoccupied Crown Lands and railway rights-of-way were exempted and became infested with rabbits. Furthermore, while government agencies have tried to have farmers construct rabbit-proof fences, closer settlement schemes have been so surveyed that boundaries cross watercourses and other natural features which have rendered permanent netting fences virtually impossible to maintain (Rolls, 1977: 97). From time to time de-commercialisation, prohibiting the sale of the skins and flesh of wild rabbits, has been advocated and legislated. But the administrative and constitutional difficulties of widespread enforcement of de-commercialisation are considerable (Sawyer, 1962: 255–258). By and large, it has to be concluded that attempts to control pests by legislation have not been successful in Australia.

Mining

No sooner had squatters achieved security over the pastoral areas through Earl Grey in 1847 than the Australian colonies' experienced the bonanza of the gold rushes. The gold rushes were to have far reaching consequences, among which were: rapid population growth; a thorough technological transformation of the pastoral industry; universal manhood suffrage; expansion of infrastructure, in particular transport and communications; large influxes of British capital; and the stimulation of a diverse range of industry. Short term social disruption and environmental damage was a small price to pay for the long term benefits of the mining boom.

The gold rushes of the 'fifties in New South Wales and Victoria attracted a flood of immigrants from all over the world. In ten years the population increased from 400,000 to 1.1 million (Maddox and McLean, 1987: 10). This sudden influx of people created a demand for foodstuffs which encouraged pastoralists who had lost shepherds to the goldfields, to concentrate on producing meat for the mining population, while the demand for breadstuffs on the goldfields suddenly created a flourishing export market for South Australian wheat. At the same time developments overseas were opening up the prospects for Australian grain to enter the international wheat trade. The repeal of the corn laws in 1846 opened up the English market to cheaper foreign suppliers. The increase in the world's shipping capacity and the linking up of the continents by submarine cable together with the increasing dependence of European markets on overseas sources contributed to the development of international markets for foodstuffs.

Mining greatly increased the colonies' ability to attract overseas funds. Attitudes of British investors to Australia changed considerably. About two million per annum of capital was imported into Victoria and New South Wales between 1827 and 1850, compared to about 340,000 between 1827 and 1850 (Doran, 1984). The capital inflow provided the funds for the rapid development of the railways in the 1850s and 1860s and the extension of telegraph and communications from the 1860s to the 1900s. Funds also flowed into residential construction and the pastoral industry. Doran concludes that

Immediate dislocative effects proved transient and insignificant compared with the long term benefits. (Doran 1984: 50)

Whereas the gold rushes in California were virtually unregulated, in Australia a degree of regulation was imposed through the insistence that every digger purchase a licence. Invoking a sixteenth century lawsuit, Governor FitzRoy pronounced the Crown's right to all gold found in New South Wales. These licences served two functions. They provided the government with money to police the goldfields, and were intended to discourage at least some of the would be diggers from joining the rushes. Furthermore, by restricting each lease to a surface area of eight feet square, the regulations increased and concentrated the population on the diggings thus exacerbating the environmental damage of mining (Blainey, 1963: 20–21, 32).

Despite mining operations being so widespread over the face of Australia during the nineteenth century, severe degradation of the environment was generally confined to the specific localities in which mining took place. In the case of alluvial gold mining in the 1850s in Victoria and New South Wales the environmental impact was intensified by the huge concentrations of population which were attracted to the diggings.

Mining operations made great demands on water and timber resources and miners were allowed to take control of these resources in their localities. The major environmental impacts of mining were therefore not directly related to the mines themselves.

Wooded areas in the vicinity of the minefields were clear-felled as miners scavenged for timber to cut into slabs to line their shafts or props to support tunnels. Bark was stripped from stringybark trees for roofing, walls and gutters. The trees were effectively ringbarked and, when dead, cut for fuel. In the 1840s timber in the vicinity of the copper mines at Kapunda, Burra and Wallaroo in South Australia was soon stripped and within a few years mallee was being hauled from 20 or 30 miles away to fuel the smelters which consumed up to 150 tons of wood daily. In the arid areas around Broken Hill and Coolgardie the denudation of the sparse timber was even more complete. Even in the heavily forested areas around Zeehan, Mt Lyell and Mt Bischoff in Tasmania the timber resources were depleted at an alarming rate (Williams, 1975: 122–23). So heavy was the demand for timber in the Victorian mines, that trees were cut in Tasmania and shipped across Bass Strait. The traffic in timber from the Otways in southern Victoria to the gold fields, prior to 1930, was so great that it used to be said that there was more Otways timber below ground than above it (Houghton, 1975: 37).

Lack of clear legislative specification regarding water rights for mining provided opportunists with the chance to build water races and sell the water to miners. This focused attention on the issue of appropriate water provision for mining communities. In Victoria in the 1860s, legislation was passed to more clearly define licences for races and reservoirs in order to preserve the rights of mining communities as a whole and also to provide subsidies to encourage the construction of water storage and reticulation systems by local authorities. This established the precedent for public ownership of key resources (Powell, 1975: 39).

To some extent the mining operations themselves were disruptive to the environment through excavation, the dumping of earth and tunnelling. Mullock heaps, often contaminated with mine or smelter residues, provided a rooting habitat foreign to most native species. Introduced species have been widely used to stabilise areas disturbed by mining. The same is true for coastal dunes damaged by sand mining. The noxious South African shrub, boneseed, has largely replaced the native coast wattle (Wace, 1985: 145). Sludge from mining operations sometimes blocked the natural drainage patterns exposing agricultural and residential land to inundation. In the north-east of Victoria hydraulic mining was introduced by Californian miners and caused the undermining of hillsides and the banks of streams, exposing the bed-rock. Deep sinking and tunnelling, which were introduced in Victoria as the alluvial gold ran out in the late 1850s and 1860s, disturbed local water tables and were responsible for large mullock heaps (Powell, 1975: 37–39).

Many of the effects of mining on the environment were only temporary. In time the cut-over forests and woodlands grew back to second and even third generation re-growth. While evidence of mining operations can still be seen in abandoned areas—these tend to be localised effects which must be balanced against the long term benefits of early mining operations. It must also be recognised that the environmental impact of gold mining in the nineteenth century was much greater than modern open-cut mining. Reconstruction of the surface following mining has become standard practice and slopes are regraded and vegetation replanted with the advice of environmental scientists (Heathcote, 1975: 134). The pioneering regeneration of Broken Hill by the mining companies in the 1930s has been repeated at many mine sites throughout the continent with the consequence that the environment is now more suitable for human habitation than it was before mining (Kearns, 1982: 140). However, the costs of this form of rehabilitation should be balanced against its benefits.

Encouraging agriculture

With the decline of alluvial mining from the 1860s, increasing numbers of ex-miners found themselves without a stake in the country. In a society where ownership of land seemed to be the path to sturdy independence they demanded that the land, so recently put into the hands of the pastoralists, be made available to them. These demands were reinforced by the power of the vote and the widespread belief at the time that small scale farmers were the salt of the earth in a democracy. This strong community belief in the virtues of small scale husbandry plus the burgeoning market opportunities, created pressures on government to establish a yeoman farming settlement (Heathcote, 1982: 108).

The first attempts to draft legislation to expropriate pastoral leasehold land and convert it into freehold for small farmers in the sixties failed. Pastoralists turned key sites like river frontages and water holes into freehold thus managing to retain control of their estates. However, legislative efforts, from the late 1870s on, did wrest the land from the pastoralists and hand it over to family farmers through the passing of Land Acts (Roberts, 1968: Chaps 19,20). The legislation not only failed to recognise established property right conventions but, as is inevitable with much centralised decision making, it had unforeseen consequences.

A feature of these state sponsored closer settler schemes was that the subdivisions under particular acts were of uniform size, irrespective of the terrain, the quality of the soil, or the nature of vegetative cover. They also usually assumed that the farmer would make his living from cultivation rather than by grazing, except of course in the case of dairy farming. The drafters of legislation invariably sought to achieve maximum intensity of settlement as this met the bureaucrat's criterion for success—the number of farmers settled in a given area. Intensification of settlement was also meant to lead to more capital investment and hence higher production and the conserving of costs in the provision of infrastructure. In many cases, however, settlement under the land legislation resulted in farmers being placed on inappropriately sized blocks for the location. In the course of time, the unsuccessful settlers moved on and their farms were incorporated into those of their more successful neighbours. The same process of aggregation of holdings also accommodated the need for larger holdings, especially in the wheat growing areas, as new technology increased the economically optimal acreage.

A consequence of the commitment by government to the societal goals of closer settlement was the obligation of government to provide support for settlers once they were on the ground. As settlement spread further from the coastal area, transport became a costly impediment. Governments were called upon to provide railway communication and, when settlement spread beyond twenty miles of the railhead, further extensions were called for. When yields fell or prices dropped, governments were expected to lower the rail freights.

As Powell notes, 'The role of the state had been ambitiously proclaimed as creator of landed opportunities, arbiter of land quality, watchdog for the public interest, architect of a new society' (Powell, 1988: 17). In reality government intervention in the process of land settlement provided a crutch to environmentally and economically unviable settlement patterns. The experience of coming to terms with the Australian environment was a difficult one, but when governments either actively assisted settlers, or picked up the tab for costly mistakes and errors of judgement, the pain of adjustment was prolonged.

Not all government efforts to encourage the development of agriculture were counter productive. The establishment of the various agricultural and lands departments in the colonies assisted farmers in adapting to the environment. Lands department surveyors made assessments of the county at the margins of settlement which gave some indication of the productive use to which the country might be put. Perhaps the most famous of these assessments was that made by the South Australian Surveyor General, George Goyder, in the drought of 1865. Goyder's 'Line of Rainfall' (which was in reality a vegetation line marking the southern boundary of the mallee, salt-bush and blue-bush country, but which was later found to coincide with the 12 inch isohyet), was used to demark the country suitable for agriculture from that suitable for grazing. Although with good seasons in the early 1870s agricultural settlement advanced beyond Goyder's Line, a decade later the wheat frontier had retreated back behind the fringes of the Mallee country (Meinig, 1963: 91-92; Price, 1966: 157).

Although the Mallee country was later to be conquered for grain farming, the work of Departments of Agriculture experimental stations on dry farming techniques together with the development of suitable crop varieties for particular localities helped make permanent cultivation of the semi-arid interior possible. And when cultivation took its toll in soil erosion, the government sponsored Soil Conservation Authorities providing the expertise to assist in tackling the problem.

Wind erosion had been a problem in the light rainfall areas of South Australia, Victoria and New South Wales since the early part of the twentieth century. The first blackout dust storm hit the Mallee in 1902 but it was a long time before official action was taken to combat the problem. A Soil Drift Act was passed in South Australia in 1923 and a Soil Conservation Committee thoroughly investigated the problem and reported in 1937. A Soil Conservation Service was established in New South Wales in 1938 while Victoria set up a Soil Conservation Board in 1940 (Dingle, 1984: 188, 245). These various soil conservation services largely performed an educative function by advising farmers on ways to combat the problems of erosion on their properties and generally raising community consciousness of the importance of soil conservation.

Soldier settlement

Australia has always been a highly urbanised society, but while most Australians chose urban living for themselves, they tend to lament the fact that so few others chose the rural life. Closely settled rural communities were perceived as model environments in which to raise strong, independently-minded, loyal citizens. They were seen as offering a counterbalance to the corrupting influences of urban life and the artificial values of industrial society. It should not, therefore, be surprising that proposals to settle returning soldiers from the Great War on the land received public approval as a way of repaying the 'debt of honour'.

In spite of a wealth of previous experience with closer settlement, the administration of the soldier settlement schemes managed to ignore most of the lessons which could have been learnt from the past. Only Western Australia and Queensland had sufficient suitable areas of Crown lands to devote to soldier settlement. The other states had to resort to heavy government expenditure on land purchases.

It was not only the high costs of establishment borne by the settlers and the impact of low prices during the Depression which were responsible for the failure of the scheme. Factors such as the poor location of farms, and inappropriately sized farms for particular localities, as well as the high capital debt burdens assumed by the settlers, and the unsuitability for farming of many of them—all contributed to failure as they had to many earlier closer settlement schemes (Powell, 1988: 100–110).

Irrigation

Belief in the societal virtue of intensive cultivation, even in spite of the evidence that this is only rarely compatible with the Australian environment, accounts for the immense public support for irrigation schemes in Australia. The stark dryness of the Australian interior inspired the dream of arid areas transformed into green oases.

As early as 1835 Wakefield had suggested that irrigation would be necessary in South Australia to provide insurance against drought (Martin, 1955: 19). Demand for irrigation received some impetus in Victoria after the 1865 Land Act's success in enabling agricultural settlement on the Northern Plains. After a run of good seasons in the first half of the 1870s drought struck in 1876–77 and the settlers on the Northern Plains looked to the Government to assist them by providing irrigation facilities.

Irrigation came to be seen as the panacea which would realise the dream of a citizenry of sturdy yeomen farmers. Many schemes were proposed. Perhaps the most fantastic was the brain child of Mr Benjamin Dods who in 1871 issued a prospectus for the Grand Victorian North-Western Canal Company. This contained a proposal for a private company to construct a canal on land grant principles across northern Victoria and then south to the sea at Portland. The canal was to be not only for irrigation but also for navigation (Powell, 1989: 84–90).

Although not a feasible proposition, the Grand North-West Canal scheme captured the imagination of the public and helped fuel the fervour for irrigation which culminated in a series of reports on irrigation in the early years of the 1880s and the Royal Commission on Water Supply which reported in 1885 and 1886. So great was the interest in irrigation that nearly every year between 1880 and 1891 saw the publication of the report of one official enquiry or another (Martin, 1955: 26). Alfred Deakin, the Minister for Water supply in Victoria, undertook an extensive study of irrigation in the United States which was published as *Irrigation in Western America*.

Deakin encouraged the Chaffey brothers, who had pioneered successful irrigation colonies in California, to come to Australia where they established private colonies on land supplied by the government at Mildura in Victoria and Renmark in South Australia (Fogarty, 1967). Most of the recommendations of the Royal Commission on Water Supply were implemented in the Irrigation Act of 1886. The main provisions of this act were the nationalisation of water use rights, licensing of private diversions from the aqueduct, state construction of irrigation facilities and the authorisation of loans to the irrigation trusts (Powell, 1989: 112).

Perhaps the most significant aspect of the legislation was the nationalisation of water use rights. This did away with riparian rights which were inherited from England under common law and replaced it with state ownership of water use rights. The approach was later to be adopted by other Colonies and States. The irrigation trusts accumulated large debts to the Government and were abolished by the Water Act of 1905 and replaced by a central agency, the State Rivers and Water Supply Commission. A second feature of this legislation was the nationalisation of the banks and beds of all watercourses (Powell, 1989: 147).

Most of the irrigation schemes in Australia are located in the Murray Valley. The most significant irrigation area outside of Victoria is the Murrumbidgee Irrigation Area in New South Wales which was officially opened in 1912 (Langford-Smith & Rutherford, 1966: 33). In more recent times a number of large scale irrigation schemes have been undertaken in Northern Australia, including the Ord River scheme in Western Australia and the Burdekin Falls Dam near Townsville in Queensland. Although such schemes may enhance the national pride of the citizens in the States where they are located, there seems to be little indication that the politically minded promoters of modern irrigation schemes have learnt much from the economic mistakes of their predecessors.

Keith Campbell observed that, 'water and irrigation seem to bring out all that is irrational in man' (Campbell, 1980: 141). Perhaps one of the most irrational aspects of irrigation policy has been the under-pricing of water to farmers. The costs of dams and other headworks has been traditionally borne by the state rather than by the irrigators. This has not only led to the uneconomic use of water and raised unrealistic expectations about investment in irrigation compared with dry-land development, but also to a more profligate use of water which over time has contributed to the huge problem of salinity, particularly in the Murray Valley.

The prospect of converting the great arid expanses of inland Australia into blooming gardens has been one of the persistent dreams of Australians. This dream was accomplished with the construction of the Snowy Mountains Scheme in the immediate post-World-War II period. This scheme diverted the waters of the south flowing Snowy River northwards into the Murray and Murrumbidgee Rivers, but perhaps the grandest dream of all has been that of diverting water from the heavily watered eastern side of the Great Dividing Range in Northern Queensland west into the headwaters of the Thompson River which, as Cooper Creek, flows into Lake Eyre. First suggested by Dr J.Bradfield in the 1930s, this scheme is unlikely to be economically feasible, yet it is periodically resurrected, usually in election campaigns (Davidson, 1969: 227–230).

Managing the urban environment

A high degree of urbanisation is a notable characteristic of staple export economies. The production function of the export economy requires a concentration of economic activity around the ports to provide services like transport, stevedoring, ship chandlering and financial and insurance functions. In addition to the export services, importers established merchanting and fabricating establishments and the main port cities became the administrative centres for the colonies. The urban development associated with the port and administrative functions formed a nucleus which attracted manufacturing and numerous service industries to meet the needs of the urban population.

Concentrations of population give rise to problems of effluent disposal, both private and industrial. Access to water and air resources for waste disposal is difficult to restrict. The common pool characteristics of waste disposal require a level of government control so that overall limits can be imposed once congestion becomes a problem within these boundaries. Private sector initiatives can offer efficient waste disposal services. However, in Australia the tradition has generally been to solve these problems by regulation and by government provision of infrastructure and essential utilities. From the beginning, the control of settlement included waste disposal.

From the earliest settlement at Sydney Cove the authorities took measures to ensure a public water supply. This was achieved by deepening the river and excavating tanks in its sandstone bed. Thereafter land use was directed so as to preserve the Tank Stream from serious pollution and erosion and regulations were passed forbidding the felling of trees within fifty feet of the stream and fences and ditches were placed on each bank for further protection. In 1802 Governor King proclaimed that vigorous action would be taken against persons who discharged noxious effluents into the stream (Powell, 1975: 55).

As in other parts of the world, the growth of cities in Australia led to the imposition of regulations to control effluents and the construction of infrastructure to provide for the storage and reticulation of water and the carrying away of pollutants. Melbourne provides an example of the problems of managing a quickly growing urban environment. Melbourne grew very rapidly following the gold rushes of the 1850s and, although the centre of the city contained many fine public buildings and was well serviced, the areas adjacent to the rivers attracted heavy industries which discharged their wastes into the waterways without restriction (Dunstan, 1984: 136).

Concerns about sanitation and pollution tended to be ignored until either the accumulation of sewage, garbage and decomposing animal carcasses became unbearably offensive, or there was a scare about epidemics. The usual response was to try to shift the problem into someone else's backyard. Garbage disposal was a problem and people who lived in low lying areas were likely to be the recipients of sewage and garbage from the higher areas. In 1854 anti-pollution legislation provided penalties for tanners and wool scourers who discharged effluent into the Yarra. This legislation met with bitter opposition from local businessmen who mounted a campaign to 'unlock the Yarra for the peoples' factories' (Barret, 1971: 45, 72, 106). A full scale 'environmental' debate ensued with the 'developers' insisting that the law was an unwarranted attack on legitimate employment-generating industry. On the other side were those who insisted that the law did not go far enough, jeopardising the health of the citizens.

The dispersion of powers amongst various authorities; the Colonial government and the various municipal authorities; made it difficult to satisfactorily resolve conflicts of interest or to co-ordinate the provision of services like sewerage and water supply. The disposal of nightsoil became a pressing problem with the spread of the urban area. Manure depots on the fringes of settlement were established and in the 1860s night soil collected in the Fitzroy and Collingwood areas was emptied into the Yarra. As early as 1855 a special rate levied by municipalities for collection of night soil was suggested, although such a special rate was not made legal until 1890 (Barret, 1971: 72, 79).

Lack of adequate sewage disposal and a contaminated water supply took a heavy toll on the health of the people of Melbourne over the four prosperous decades from the gold rushes of the fifties to the depression of the nineties. Melbourne's infants were decimated in periodic waves of enteric disorders, diphtheria, scarlet fever, measles and whooping cough for want of a basic infrastructure for water, drainage and sewage (Dunstan, 1984: 125).

Measures to provide a reticulated water supply did not always solve the problem of contaminated water. The Yan Yean system was turned on at the end of 1857 but the reticulation included street hydrants which let the air in. At times of low pressure they would allow in street matter, creating a health hazard. In addition the reticulated water supply encouraged many to illegally install water closets which added to the contamination of ground water and spilled over to the low lying areas (Dunstan, 1984: 138). Melbourne may have been a distinguished Victorian city but its most distinctive characteristic throughout the Victorian era was its smell. As the Age commented, 'Stink has become quite a Melbourne institution' (Dunstan, 1984: 251). Yet, in spite of the dangers to health and the affront to sensibility, it was not until 1889 that the political will could be mustered to create the Metropolitan Board of Works, which was to undertake the construction of an efficient sewerage system and to attend to the problems of drainage and water supply.

Conclusion

Economic activity inevitably has an impact on the existing natural environment. While resources are abundant, productive management of the environment is a rational choice. The early development of Australia's fishing, timber, agricultural and mining resources, laid the foundation for her future prosperity. It was inevitable that in the early stages of settlement ignorance about the ecology and poorly defined property rights would result in some damage to the environment. But market forces, evolving systems of land tenure, and better information would have led to a greater emphasis on conservation and preservation as Australia developed. Nevertheless, from the earliest days of European settlement, governments undertook the major responsibility for the management of the environment in Australia.

Although the economic development of the continent was driven by market forces, governments have been reluctant to use market mechanisms in managing the environment. Overwhelmingly, regulation has been the preferred instrument for environmental management. This strong reliance on regulation has been attended by some unforeseen consequences.

Governments became heavily involved in the provision of infrastructure, and the political pressures often resulted in distortions in the uses of resources which actually contributed to environmental damage, rather than ameliorating it. The provision of irrigation water at considerably less than cost, for example, has massively contributed to the problems of salinity in the Murray Valley. Heavily subsidised rail freight rates have encouraged farming in fragile environments while unrealistic drought relief schemes have encouraged inappropriate stocking rates in low rainfall areas.

Closer settlement schemes were often devised with more regard to bureaucratic convenience than to the limitations of the environment. Regulations rarely took account of the differing nature of the land, and farm areas were often quite inappropriate to the region with the consequence that inappropriate land use resulted in erosion problems. In other instances lack of adequate information regarding the potential of the environment led to bad judgements regarding the appropriate land use. Although in a less regulated system these errors of judgement would still undoubtedly have been made, the inflexibility of the regulatory system compounded the mistakes and intensified the deleterious consequences.

Governments are rather inclined to seek solutions in regulations but are less enthusiastic about providing the resources to monitor and enforce the regulations. In some cases, as with regulations compelling landholders to exterminate vermin, the task of monitoring and enforcement is very difficult and almost impossible when government itself does not feel obliged to eradicate pests on Crown lands.

Evaluated from the stand-point of today's preference for conservation and preservation and today's superior knowledge of ecology and ecosystems, both markets and governments have failed to optimally manage environmental resources. However it is unlikely, given a similar information set and similar pressures to make a living from the land, that we would have acted differently. Market forces may have cut down forests, opened gold mines, and created environmental change. But they also created future opportunities for more efficient resource use and a higher standard of living which now enables choices towards conservation to be made. Rather than lament the past, it is more appropriate to ask what are the most effective mechanisms for managing the environment today? Given that both markets and governments fail, how we can learn from the lessons of the past and utilise the comparative advantage of each institution to balance the trade-off between productive use and preservation.

4 Land degradation

Executive summary

Land degradation refers to the adverse effects of commercial agriculture on future levels of agricultural production and on the supply of other environmental services. Decisions on grazing levels, on cropping practices, on irrigation and on clearing land—all designed to increase production and income—can result in soil erosion, loss of soil fertility, salinity and other forms of land degradation. Clearly mistakes have been made in the past; there are well documented examples of serious land degradation, and there is a prima facie case that under current institutional arrangements some current farm management strategies generate too much land degradation. However, the past is not all bad; over the last forty years Australian agricultural output has trebled, and further increases are projected.

A framework for evaluating farm management decisions and government policies influencing land degradation is described. The focus is on what should be done in the future rather than on rectifying past mistakes. Answers vary according to the different types of land degradation, the areas affected, and due to changes in information about technology and market prices.

Where most of the social costs of land degradation fall on individual farms as on-site costs, individual farmers will choose roughly the correct level of land degradation. This is the case for management decisions resulting in loss of soil fertility and for most cases of soil erosion. For these circumstances, the role for governments is to avoid distorting commodity prices and the costs of inputs, to provide secure land tenure, and to provide an institutional structure for research and extension.

Land degradation in the form of salinity is a different issue. Off-site or externality costs of individual farmer irrigation and land clearing decisions may impact on the well-being of other farmers and water users. Individual farm decisions may result in excessive levels of salinity. Continued reductions in government subsidies for irrigation water and for irrigation-intensive products will help. Complexities due to the externality effects associated with salinity mean that it is not automatically valid to argue that government imposed taxes, subsidies or regulations to reduce salinity will improve national welfare.

Introduction

Recent years have witnessed an increase in concern about the quality of Australia's agricultural land base. Land is an important determinant of the current and future productive capacity of commercial agriculture and of the capacity to provide other environmental services. As noted in Chapter 3, two hundred years of European settlement has produced graphic incidences of soil and wind erosion, salinity, loss of wildlife and other forms of land degradation in many parts of the country. Underlying causes of land degradation include individual farmer error in applying European technology to a different environment, and government policies for closer settlement. The media find land degradation good copy. Several government reports (for example, Department of Environment, Housing and Community Development, 1978) and individual studies (for example, Eckersley, 1989) claim land degradation is costing Australia upwards of $2 billion per year in lost agricultural production. However, it also has to be recognised that agriculture output has risen nearly threefold over the last forty years. Since bygones are bygones, the relevant question is whether future cropping, grazing, irrigation and other decisions of individual farmers and the policies of governments influencing those farm management decisions affecting future levels of land degradation should be altered.

Land degradation is high on the political agenda with the 1990s being declared the Decade of Land Care. A fascinating point of political consensus between those for economic development and those for preservation of the environment was reached in 1989 with agreement between the National Farmers' Federation and the Australian Conservation Foundation to '.... work together towards ensuring that Australia's agricultural and pastoral lands are used within their capability by the year 2000 and that there is a sustainable use of lands from that time on.' Such warm sounding sentiments raise the questions: what is the meaning of the phrase 'sustainable growth' and how best can we meet this goal?

This chapter seeks to evaluate current farm management decisions and government policies affecting land degradation. The next section provides a background description of the nature and causes of land degradation. Then follows comments on available estimates of the costs of land degradation, and on the issue of sustainable agriculture. Building on the general framework for evaluating the use of resources of Chapter 2, we provide an economic framework in which individual farmer decisions and government policies towards the utilisation of natural resources for commercial agriculture can be assessed. The chapter concludes with a discussion of desirable government strategies influencing commercial agriculture and its use of scarce natural resources.

What is land degradation?

A broad definition of land degradation is the adverse effects which various uses of land by man have on current and future services provided by land. Our interest is in the effects of commercial agricultural decisions concerning clearing land, cropping practices, grazing levels, irrigation, structural land works, and so forth. In the initial period these decisions influence output levels of wool, wheat, fruit, etc., input costs, and current cash flows and income. But also, these decisions can influence soil fertility, soil erosion, salinity of land and water, weed invasion and other characteristics of land quality. These land quality characteristics, in turn, influence future levels of commercial agricultural output. Ultimately, the principal cost of land degradation shows up as a fall in future levels of agricultural output and a fall in the future value of other environmental services such as water quality and amenity.

Land degradation can take many forms, it can involve a wide geographic coverage, and it can involve long time spans. For the purpose of evaluating the best use of scarce agricultural land from

a national perspective, it is helpful to have a picture of :
- the management strategies determining land degradation;
- the characteristics of the degradation in terms of the relative importance of on-site and off-site effects (that is, losses of land services directly falling on the individual farmer and those losses falling on others in the community). Where there are off-site or externality effects the distinction between point and non-point forms (that is, off-site effects relatively easily measured at a single point for each farm, and those effects occurring at a great many difficult and costly to measure points);
- the level of knowledge about key physical and biological relationships linking commercial agricultural management practices and land degradation.

These characteristics, especially the relative importance of on-site and off-site forms of land degradation, are important in assessing the role for and the form of government policies towards land degradation.

Loss of soil fertility is a good example of soil degradation which is dominated by on-site effects. Cropping, livestock and other practices causing soil nutrient loss (primarily via the sale of products, soil structure decline and compaction, soil acidification, water repellence and so on), reduce the future production capacity of the individual farm. That is, the individual farmer bears most of the national costs of lower soil fertility as a loss in his future income. In many cases knowledge about the key physical and biological processes behind the fertility loss will be imperfect, not only for the farmer affected, but also for other farmers and for government officials.

Soil erosion is thought to be the most pervasive form of land degradation in Australia. Shifting of topsoil by wind and water is hastened by management practices which reduce the volume of tree, shrub and pasture coverage. Land clearing, cultivation and heavy grazing often lead to soil erosion. Most of the actual erosion occurs infrequently in times of severe seasonal conditions, especially drought and flood, rather than regularly. Because of the enormous diversity of soils, landscapes, foliage, climatic conditions, etc.; the form and magnitude of soil erosion is highly variable across farms and even across paddocks of each farm. Changes in farm management practices and land structural works seem capable of partially, if not fully, reversing most cases of soil erosion. In some cases man has adopted practices which actually reduce the natural rate of soil erosion.

Most of the adverse effects of soil erosion are on-site or internal to each farmer, but there are some off-site effects. A large part of the land degradation cost of soil erosion shows up as a loss of fertility and of future productive capacity for the individual farmer. In some cases, particularly in the intensive cropping areas, soil erosion dumps unwanted material on neighbouring farms, on local roads and into local water courses. This form of externality often involves a small number of economic agents, and it is relatively easy and low cost to see, measure and monitor. In these circumstances negotiation costs often are low enough to enable the polluter and the polluted to reach a mutually beneficial agreement which effectively internalises the externalities. There are some instances in which soil erosion will have diverse and widespread off-site effects, but these cases are relatively rare. Examples include dust storms and contribution to flooding and flood damage. While there is no hard data on the relative importance of the off-site effects of soil erosion, logic and anecdotal evidence indicates they are a small part of the total social cost of soil erosion. In most cases this is less than 20%, and then the adverse effects are usually concentrated on just a few neighbours.

Salinity, and to a lesser extent water logging, has been the topic of much media attention. Irrigation is the cause of wet land salinity and water logging. An undesirable side effect of much irrigation is an increased intake of water by the soil which in turn raises the water table and/or increases the leaching rate of salt (and other 'bads'). Often this results in salt damage to farmland downstream and the salting of water in streams and rivers used by other farmers and for domestic and commercial use. In the case of dry land salinity, clearing deep rooted plants, especially trees, on the slopes is a cause of increased water percolation, injection to water tables and increased levels of salts (and other 'bads') in streams and rivers. Despite the media attention given to salinity, within the wider picture of Australian agriculture it is quite small. Data for the late 1970s indicates that salinity is a problem for less than 1% of agricultural land (Shaw and Hughes, 1981, quoted in Hodge, (1982); chapter by Charles in Chisholm and Dumsday, 1987; Greig and Devonshire, 1981), although there are estimates that the area affected is increasing (for example, *The Age*, July 21, 1990, suggest the area could treble over the next 30 years). Irrigated agriculture accounts for less than 2% of Australian agricultural land and about 12% of the value of all Australian agricultural output.

The adverse effects of land degradation associated with salinity are characterised by the importance of off-site or externality costs relative to on-site or internal costs—a marked contrast to the soil fertility and soil erosion stories. Particularly in the case of dry land salinity, few of the adverse effects of tree clearing will appear on the tree clearer's farm, or even on neighbouring farms. Rather, the problems of higher water tables and salinisation are felt by farmers many kilometres away, and by other water users. For example, the clearing of trees on the slopes of the Goulburn River catchment areas in Victoria is thought to have aggravated or contributed to salinity and water logging problems for farmers along the Goulburn and Murray River flood plains and to urban users of water from these rivers through to Adelaide. Further, the time lag between the tree clearing activities and the adverse land degradation costs may involve several decades.

Similarly, although not to the same extent, a majority share of the adverse effects on the water table and salinity levels, caused by a particular farmer irrigating, are borne not by the irrigator but by downstream farmers and other water users. No doubt the relative importance of off-site costs in total social costs varies widely from situation to situation. While there are no hard numbers on the cost breakdown at an average or aggregate level, available evidence suggests that for many irrigation farms the off-site land degradation costs exceed half of the social costs (see, for example, the papers in Taylor, Dumsday and Bruyn, 1982).

Formulating government policies for the salinity problem is

particularly difficult for reasons additional to the complexities generated by off-site effects. There is limited information about the complex physical relationships linking managerial decisions on land clearing and irrigation to the ensuing land degradation problems. Although working computer models purport to explain key relationships (for example, models used by the CSIRO and described in Brett, 1990) they rest on contentious assumptions and parameters which significantly influence model outputs. Most off-site salinity can be characterised in terms of non-point externalities. The effect of this is that it is very difficult and/or costly, (if not impossible), under current technology to measure and to monitor the contribution of different management strategies to the off-site costs of salinity.

Land degradation embraces a range of measures for reducing in the future productive capacity of agricultural land by current farm management practices. The heterogeneity of the different forms of degradation has important implications for answering our questions: is there too much land degradation, and what policy changes are required to achieve desired changes in farm management decisions?

Land degradation and sustainable growth

While a number of estimates have been made of the extent of land degradation in Australia, all can be criticised. In particular, the goal of restoration to some pristine state is at variance with the goal of sustaining growth in living standards. In the context of sustaining the productive capacity of Australian agriculture, numbers on output and inputs for the past forty years clearly indicate an ever expanding productive capacity.

Decisions by the commercial agricultural sector on clearing, cropping, grazing, irrigation, etc. as well as to degrade land and to restore degraded land—are just a subset of the general problem of how best to utilise limited resources to meet the insatiable needs of society. The needs of society are taken to embrace not just today's needs but also the needs of tomorrow and the day after that; and they are taken to embrace non-market services such as genetic diversity and aesthetic appeal as well as market sales of food and fibre. Using land in one way, such as arable grazing, has an opportunity cost, such as using the land for irrigated horticulture, as well as land degradation costs. Again, using capital, labour and other resources to repair soil erosion means those resources are not available for other productive uses, such as cropping and medical services. Whether or not land degradation is a real problem can only be determined once we have found out whether a change in current agricultural practices influencing degradation adds more to social well-being than the opportunity cost of those changed practices.

Most available estimates of the extension and costs of land degradation use the pristine state or natural degradation rate as the point of reference. That is, they try to estimate the extent of soil erosion, salinity, loss of soil fertility and so forth relative to the state of the land before the onset of commercial agriculture, and/or they try to estimate the costs of restoring land to this pristine state. Even if this technical reference point is accepted, there is no agreed methodology and not surprisingly a wide range of estimates have been reported (Milton et al., 1989). One of the most widely quoted and respected estimates of the extent of land degradation, the report of the Commonwealth-State Collaborative Soil Conservation Study published in 1978 (with results described in Woods, 1984) estimated that 55% of Australia's arid areas and 45% of its non-arid areas are subject to soil erosion, and that structural works of up to $2 billion (in 1990 dollars) would be required to correct the degradation. Others have estimated that engineering works costing 100s of millions and even billions are required to reverse salinity problems (papers in Taylor, et al., 1982; and *The Age,* July 21, 1990).

While these technological referenced estimates may be of interest as measures of the extent of land degradation, they do not tell us whether land degradation in Australia is excessive or not nor whether there exists a problem to be solved. The relevant question is whether Australia as a nation would gain by adopting different farm management practices with regard to land clearing—cropping, grazing, irrigation, land structural works, and so forth—which would alter or even reverse land degradation. The reality is that past mistakes have been made and cannot be altered.

We need to ask: Can we justify changing management decisions now and in the future recognising that scarce resources mean that (say) lower grazing rates and expenditure on structural soil conservation involves opportunity cost—that is, foregone current production and use of resources.

An understanding of the mechanics of soil degradation discussed in the previous section points to three sets of reasons for arguing that land degradation is a potential problem. The first two reasons concern the issues of market failure. First, where off-site forms of land degradation (or externalities) are important, individual farmer decisions which ignore these externality effects will not result in the best use of society's scarce resources. This seems to be a small problem for farming decisions affecting soil erosion and soil fertility. Here, what is good for the farmer is good for society. However, in the case of salinity in which off-site (or externality) problems are important, it is likely that this form of land degradation is excessive. Using an illustrative model, Quiggin (1988) estimates that a shift from current irrigation strategies on the Murray River basin to a socially optimum strategy could raise net value added by this region by 30%.

Second, apart from these externality problems, the decisions of individual farmers may not be in the best interests of society because their management choices are distorted by market prices for inputs, for example water and fertiliser, and for outputs which do not reflect social opportunity values, and by information deficiencies. In practice, such distortions are important in Australian agriculture; and we discuss them in detail below. However, the net effect of these price distortions on land degradation is not always clear.

A third potential source of causes of a misallocation of resources to land degradation derives from the theory of the private interests for regulation. Here it might be argued that particular pressure groups operating through the political system, such as the National Farmers' Federation and the Australian Conservation Foundation, use the land degradation issue as a cover for promoting policies which redistribute national wealth and utility to their members and in so doing misallocate the nation's allocation of resources. Whether such political behaviour has a net

Table 4.1
Measures of outputs and inputs of Australian commercial agriculture

	\multicolumn{5}{c}{Three year averages centred on}
	1951-52	1961-62	1971-72	1981-82	1988-89
Farm Output					
Index of volume of farm production (1979-80 = 100)	45	63	85	94	114
Farm Inputs					
Total area of farms (million ha)	441	473	499	490	470
Total employment ('000)	479	448	408	384	393
Real Value of Prices Received					
Index of prices received to CPI (1979-80 = 100)	89	90	86	100	99

Source: Derived from *ABARE Commodity Statistical Bulletin,* December 1989, AGPS, Canberra.

positive effect or negative effect on land degradation is not clear.

Another way of looking at the costs of land degradation in Australia is to ask whether the productive capacity of commercial agriculture has been declining. A principal concern with land degradation is that it reduces productive capacity in the future. As noted in Chapter 3 and elsewhere, there are well documented cases of land degradation throughout the history of European settlement in Australia; and inspection of the countryside will reveal particular instances of severe gully erosion, of sand halfway up fence lines, of salt patches with dead trees, and so forth. But such a country inspection would also reveal structural works to reverse soil erosion, and introduced pasture species and fertilisers which have increased soil fertility and which have reduced natural land degradation rates. In addition, new information and technology has enabled commercial agriculture to increase current and future productive capacity.

In terms of an overall measure of productive capacity it is clear that Australian agricultural output has increased dramatically, and that further increases in the future are anticipated.

Table 4.1 provides some measures of farm output, of the inputs used in producing that output, and of the real price received for farm products over the last forty years. Production has risen by nearly a factor of three, or about 2.5% per annum. This increased output has been produced from about the same area of land, with less labour, and with more capital and technological expertise. Further, ABARE in its longer term projections anticipates further increases in real output. From the perspective of sustainable development, commercial agriculture today and in the near future is far more productive than it was ten, twenty, forty and more years ago.

An economic decision making framework

Economic models of management decisions regarding the use of commercial agricultural land provide a useful framework for pinpointing the policy context and appropriate policies towards land degradation. Management decisions concerning clearing land, cropping, grazing intensity, irrigation, fertilisers, mechanical soil conservation measures, drainage, etc.—are determined to maximise an objective subject to a set of market and technological constraints. Dixon et al. (1989) provide a concise description of the different models and illustrate their application to several land degradation topics.

The objective can refer to that of a single farmer or to society, and it will refer to a time stream of current and future benefits and costs. Where there are no externality or off-site effects, as in the case of soil fertility loss, the objective of society and the objective of the individual farmer will coincide. However, where land degradation involves significant off-site effects, for example in the case of salinity, the individual farmer will not consider all of the benefits and costs of decisions which are important to national welfare.

Because land (and other inputs) provide services over time the objective function refers to a future stream of benefits and costs. Thus, for example, a farmer will compare one strategy that involves heavy stocking and land degradation this year and resultant lower stocking rates and lower land sale values in the future, against a strategy of a lower and more even stocking rate today and in the future with less land degradation and higher future land values. Future benefits and costs are brought to a present value number to enable comparison of the costs and benefits of the different management strategies.

In choosing management strategies, individual farmers and society face the constraints of market prices for their outputs, of market costs of purchased inputs, of given supplies of land and labour, and of physical and biological restraints. These technological restraints include links between different management strategies and land degradation, and between land degradation and future productive capacity of the farm. The reality of Australian agriculture is that there is considerable uncertainty about market prices, input costs, and many of the technological constraints.

Because of the uncertainty, farmers adopt a sequential decision making strategy. Given their current estimates of market prices, climate, the effects of management on land degradation, etc.; they make a decision for the next month or few years. This choice takes into account expected effects of the different land clearing, cultivation, irrigation, grazing intensity, options on current and future returns and costs and on land degradation. As the future unfolds, information becomes available to revise estimates of market prices, effects on land degradation, etc., and decisions themselves are revised on a rolling basis.

The economic model highlights key characteristics of the land degradation debate:

- we start from the current position;
- the optimum level of land degradation may be less than, greater than or about the same as the current rate or the 'natural rate';
- the optimum level of land degradation is likely to vary over time;
- individual farmer decisions can deviate from those which would be chosen by society.

Like it or not, we start the rest of our lives today, and this is also true of our approach to land degradation. What has been done, has been done and cannot be changed. Beginning with the current state of our stock of land and other natural resources, the issue is how should we best manage our land resources today and over the future. Thus, the historical performance of agriculture with reference to land degradation is irrelevant. What is relevant is the choice of current and future management strategies for the land in its present state.

In using the economic model to assess the choice of the level and rate of land degradation which contributes most to national welfare we take a special example. Suppose we want to determine the level of grazing intensity on arid land, with higher levels of grazing leading to higher levels of land degradation. This occurs as the result of the reduction of covering vegetation and the higher probability of soil erosion. Further, to simplify this first example, suppose that there are no off-site soil erosion effects and that market prices and costs represent social prices and costs; that is, the costs and benefits faced by the individual grazier also represent national costs and benefits. The management decision problem facing the grazier, and also society as an aggregate, is given in Chart 4.1.

The current benefits of a higher stocking rate and a greater risk to soil erosion take the form of increased current period wool production. However, in general the marginal or additional benefits of increased sheep numbers falls; hence the downward sloping marginal benefit curve for increased grazing intensity and for higher levels of land degradation in Chart 4.1. At the same time as the extra sheep increase current returns, their extra consumption of forage increases the risk of soil erosion. And, soil erosion means a combination of a lower carrying capacity in future years and a reduced resale value for the degraded land. In general, the opportunity costs of foregone future returns will increase with the level of excessive grazing at an increasing rate; hence the upward sloping marginal cost of the more land degradation curve shown in Chart 4.1. Now, with current price expectations and knowledge about the technological relationship (as is embodied in the two curves of Chart 4.1), the optimal level of land degrada-

Chart 4.1
Equilibrium level of land degradation

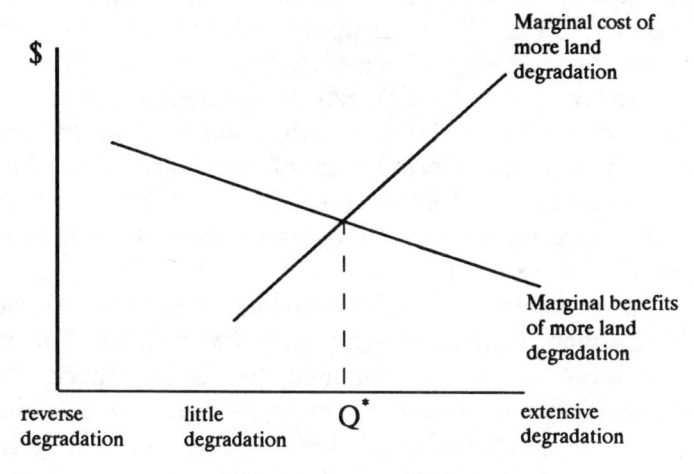

[Chart showing Marginal cost of more land degradation (upward sloping curve) intersecting Marginal benefits of more land degradation (downward sloping curve) at Q. X-axis ranges from "reverse degradation" to "little degradation" to Q* to "extensive degradation". Y-axis is $. X-axis label: Level of land degradation (due to grazing intensity)]*

tion (and stocking rate) is at Q^*. At this level of land degradation and associated grazing strategy, the marginal benefit of additional land degradation just equals the marginal cost of that extra unit of land degradation.

There is no particular requirement about the level of land degradation at which the marginal cost and benefit curves of Chart 4.1 intersect. It could be at a point of extensive degradation— almost the story of short term mining and return of the degraded land to wilderness in the future. It could be a point of reversing land degradation—for example, using structural works, planting trees and windbreaks, planting improved pastures. Only by an improbable coincidence would it be at the point of the natural rate of land degradation. This is the vital point overlooked by available estimates of the magnitude and cost of land degradation raised in the previous section.

Further, the best use of Australia's land resources is likely to mean different levels of land degradation over time. For example, changes in the understanding of the linkage between grazing and land degradation, and improved technology for restraining land degradation (for example, direct seeding versus long fallows, drip irrigation versus flood irrigation), will shift the marginal cost curve. Again, changes in commodity prices (especially of prices now relative to the future), and changes in input costs, will shift the curves. These shifts, in turn, point to different levels of management and different levels of land degradation. Australia in the past, and no doubt also in the future, has been characterised by the evolution of information about constraints, technology and prices.

The foregoing observations provide an informed framework for describing sustainable agriculture. Clearly it is a more complicated term than simply maintaining productive capacity (though the idea of maintaining productive capacity seems to underlie the goal of the joint National Farmers' Federation and Australian Conservation Foundation). In some parts of the country national welfare will be increased by a land degradation rate above the natural rate. At the same time, in other parts of the country, the sensible thing to do will involve reducing land degradation below the natural rate. Also, these rates will alter over time with the inflow of new information. Simplistic slogans about nationally desirable levels of land degradation are unlikely to be in the best interests of all Australians.

The economic model provides a convenient way of isolating underlying reasons for the management decisions of individual farmers affecting land degradation differing from those which would maximise national welfare. Individual farm decisions could result in levels of land degradation which are above or below the national best decision because:

- farmers face distorted prices for the outputs they sell and the inputs they purchase;
- farmers use a different interest rate to discount future benefits and costs than does society;
- farmers ignore a number of benefits and costs in the decision choice, in particular off-site land degradation

costs, though these benefits and costs are important in a national decision making calculus.

There are important examples in Australian agriculture where the prices received by farmers for products, and the prices they pay for purchased inputs, are significantly taxed or subsidised such that farmer prices do not represent social opportunity values. Distorted prices result in farmers increasing production of, and increasing use of, the subsidised outputs and inputs; and reducing output and inputs in the case of taxed items. Such rational, individual farmer behaviour causes an inefficient use of national resources, and often will result in land degradation levels which are too low or too high. For example, the supply of irrigation water at below cost encourages farmers not only to apply excessive water on irrigated activities, but also to choose irrigation farming in preference to dryland farming. The resulting excessive irrigation contributes to excessive levels of salinity and water logging. Similarly, until very recently, special taxation concessions for land clearing artificially encouraged this activity and contributed to land degradation. Commodity schemes generating farm prices for dairy and horticultural products above world price levels have encouraged farmers to adopt these irrigation intensive activities over dryland farming commodities. Using the economic model of Chart 4.1, any price distortions which shift the marginal benefit or marginal cost curves faced by individual farmers to the right would induce farmers to choose management strategies with higher levels of land degradation than is in the best interests of society.

Many in the environment movement make use of the argument that farmers are excessively preoccupied with current period returns, and downplay the longer term costs of their use of the land. This is the argument that individuals in their decision making use a much higher discount rate in calculating present values than does society. Choice of a discount rate is a contentious and complex issue on which there are a number of schools of thought (see, for example, the chapters by Quiggin, Kirby and Blyth, and Chisholm in Chisholm and Dumsday, 1987). At any point of time there are bound to be examples of individual farmers for short periods who are forced to use very high discount rates because of a combination of internal and external capital rationing. There farmers adopt management strategies with high rates of land degradation. However, with deregulated and competitive capital markets, and integration of the Australian capital market into the world capital market, there is a high probability that Australian farmers on average employ approximately the social discount rate in choosing decisions affecting land degradation.

The most important reason for arguing that land degradation is an economic problem of concern to all Australians is the importance of externality or off-site costs of some forms of land degradation. This is especially so where those off-site costs are of the non-point form. In the cases of land degradation taking the form of loss of soil fertility, and most of the cases of soil erosion, the off-site effects are small relative to the on-site effects. By contrast, in the case of salinity, large parts of the social costs of irrigation and land clearing occur many kilometres away, and sometimes hundreds of kilometres away. To a large extent individual farmers ignore the off-site costs in making their own decisions since they do not bear them and they are not held responsible for or accountable for them. As a result, they choose management strategies resulting in higher levels of land degradation than would be chosen by a society which takes into account both the off-site costs and the on-site costs of its decisions.

Chart 4.2 illustrates the problem where an important component of the national costs of decisions on land clearing, irrigation, etc. by the individual farmer takes the form of off-site land degradation. The Chart uses the concepts of marginal benefits and marginal costs of Chart 4.1, but here we distinguish between on-site or marginal cost to the individual farmer and to society. Individual farmer costs refer to on-site costs only. Costs to society include the sum of on-site costs and off-site costs. For example, for an irrigation farmer, the on-site costs are those of water and any salination and water logging problems caused by his irrigation on his farm, and the off-site costs are the additional salinity and water logging costs his irrigation causes other farmers and water users. Now, the farmer in making his management decisions chooses a level of land degradation at which the marginal benefits and marginal costs to him are equated, that is

Chart 4.2
Land degradation with significant off site costs

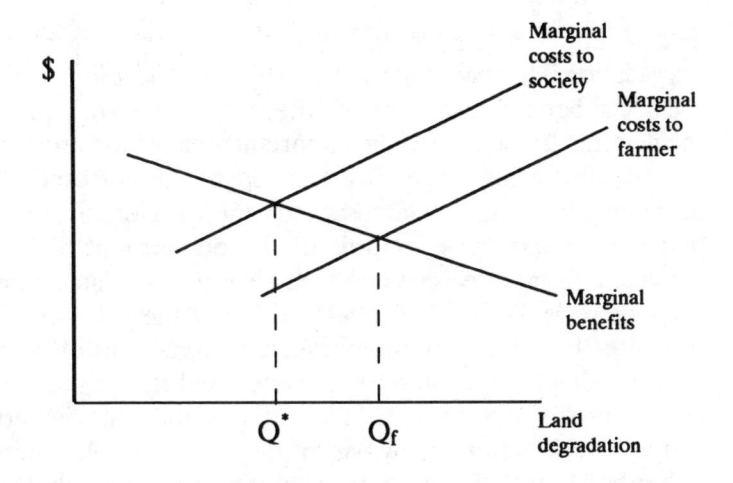

quantity Q_f. By contrast, at the social optimal where marginal benefits and social marginal costs are equated, a smaller level of degradation at Q^* is desirable. In this situation national welfare would be increased by inducing farmers to adopt management practices with a lower level of land degradation.

Demonstrating a market failure is a necessary condition to justify some form of government intervention, but it is not a sufficient condition. In reality governments are neither benevolent or omniscient. The latter is especially important in the area of salinity for the following reasons:

- the externality is of the non-point or diffuse form which is difficult to measure;
- the technological relationships between the originating farm management strategies and the externalities are complex and not well understood; and
- the magnitude of externalities generated by different farmers is likely to be highly variable.

In this situation the probability of government failure is high. Thus, there is no guarantee that government intervention will improve matters.

Roles for government

The general aim of government policies with reference to land degradation is to encourage land management decisions such that the social benefits and costs of different options are equated at the margin. In the case of the important areas of soil erosion and loss of soil fertility there is a high degree of correspondence between private, and social costs and benefits. Here the principal policy focus should be towards the encouragement of rational individual farmer decisions on land clearing, cropping, stocking and land reclamation. In the case of salinity, the policy challenge is greater because of the importance of non-point off-site soil degradation effects of individual farmer land clearing and irrigation decisions. For this type of land degradation, making private markets work better is only part of the desirable policy strategy. Other policy options—such as polluter pays taxes, establishment of common property resources, and regulations aimed at including the off-site costs as well as the on-site costs of land degradation in farmer's choice of management options—need to be evaluated.

Making markets work better

Governments can make a significant contribution to achieving nationally desirable decisions by farmers which influence land degradation, by adopting policies which reduce distortions to the incentives facing individual farmers. Particular areas concern commodity prices, input costs, land tenure, R&D, and the establishment and monitoring of property rights. These policy areas are applicable to all forms of land degradation. The aim of policies to make markets work better is to ensure that the benefits and costs used by individual farmers in choosing their management strategies are the same as the national benefits and costs of the different strategies.

Data collated by the IAC (1987) indicate significantly different rates of nominal assistance and of effective assistance to different agricultural outputs and to different agricultural production processes. Many of these different rates of assistance artificially encourage individual farm management choices, which lead to higher rates of land degradation than is socially desirable. Irrigation intensive products—including dairying and horticulture, and to a lesser extent rice and sugar—receive relatively high rates of assistance when compared with the dryland activities of broad acre farming, and grazing sheep and beef cattle. Previous subsidies for fertilisers and land clearing favoured cropping and over-grazing and, in part, they artificially stimulated expansion of wheat farming into arid areas. The highly subsidised price of water for irrigation clearly has encouraged high levels of irrigation—both in terms of an expanded area of irrigation and of higher application rates. One outcome, probably an unintended outcome, of drought assistance has been to encourage farmers to adopt higher stocking rates which in part contribute to soil erosion (Robinson, 1982; Freebairn, 1983; Milton et al., 1989; and Drought Policy Review Task Force, 1990). Acceptance of the Drought Policy Review Task Force's recommendations—in particular that of changing attitudes and policies to recognise drought as a normal, recurring climatic risk rather than a national disaster, and reducing drought assistance to meet the specific objective of welfare support—would significantly reduce current incentives to adopt management practices which increase land degradation above the socially desirable level. A policy strategy of reducing the disparate levels of industry specific assistance to different agricultural products, inputs and production processes would raise overall national productivity; and also reduce land degradation towards a more socially desirable level.

Land tenure arrangements potentially have an important bearing on farm decisions influencing levels of land degradation. Where the farmer has full ownership of land in the form of freehold or a leasehold in perpetuity, the farmer takes a long term view. If exploitative management strategies are chosen today, the individual farmer bears the full costs in terms of lower future services from land and/or a lower sale value of degraded land. Molnar (1955) and King and Sinden (1986) provide evidence of a link between farm values and land degradation. That is, secure long term land ownership means the individual farmer is responsible for the longer term preservation of soil, and he faces the benefits and costs of different management choices.

By comparison, where the land tenure system is that of short term leases, the individual farmer does not bear the full costs of land degradation. In general, limited tenure encourages the lessee to choose clearing, stocking and cultivation strategies which result in higher levels of land degradation than is socially desirable (Kirby and Blyth, 1987). While much of Australian agricultural land is held under leasehold, especially in the arid zone, often the period of lease is several decades or more and covenants are imposed on land usage; although Young (in Chisholm and Dumsday, 1987) notes that these covenants are frequently poorly monitored and infrequently enforced. There is some debate as to whether the tenure system in practice has had a significantly different effect on land degradation (see for example, the chapters by Bradsen and Fowler, Young, and Blyth and Kirby in Chisholm and Dumsday, 1987). Even granted these empirical doubts, a freehold tenure system provides a relatively low cost information system of land tenure for ensuring that individual farmers are fully responsible for, and are rewarded or penalised for, the longer term effects on land degradation of their current period management decisions.

An important influence on the effectiveness of individual farm decisions in general, and of the effects of these decisions on land degradation in particular, is the information available to the farmer about current and future prices and costs and about the effects of different decisions on land degradation. Much of this information has classic public good attributes (joint consumption and high costs of exclusion) which means if left to the private market too little resources would be devoted to research and development. There is then, a strong *prima facie* case for government intervention to increase the supply of information to farmers about the short term and long term effects of different management strategies on soil degradation and on current and future levels of costs and benefits. Milton et al. (1989) propose a number of recommendations aimed at rationalising and improving research and extension on land degradation issues.

The form which government intervention aimed at increasing research and extension about land degradation should take is more controversial. Options include direct government supply, subsidy of private research and extension, and legislation facilitating the formation and operation of rural research organisations which internalise the costs and benefits of rural research. Since the farm sector is the principal beneficiary of these research and extension activities, and because public subsidies involve considerable efficiency costs (including the deadweight costs of taxation and the deadweight costs of over-expansion of the rural sector), there is a compelling case for arrangements which collect the funds from the rural sector itself.

Where some of the costs of land degradation are off-site or externality spillovers—and particularly where these spillovers are well defined, relatively easy to monitor and affect only a small number of other farms and enterprises—government may be able to establish a system of property rights which effectively enables the spillovers to be internalised. A potential example is the case of water caused soil erosion in which run-off sedimentation from one farm adversely affects a few neighbourhood farms and/or a local road and water storage. The Victorian LandCare program is an example of a mechanism for facilitating the adoption of a common property solution to land degradation problems confronting a small group of farmers.

Unfortunately, matters are more complex with reference to the land degradation problem of salinity where off-site costs are an important component of total land degradation costs. The externality effect of tree clearing in the case of dryland salinity and of irrigation in the case of wetland salinity is characterised by: lack of knowledge of the key physical relationships; the non-point or diffuse nature of the externality; the large number of other farmers and other water users adversely affected; and measurement problems. With current technology and knowledge it is not possible to envisage a market solution.

Policy options for the salinity problem

A number of potential instruments might be considered for inducing farmers to change management practices in ways which reduce salinity. These include: a polluter pays tax on irrigation and land clearing activities which lead to salinity (OECD, 1975, and Dumsday, 1983); subsidies for activities which reduce salinity and water logging (Balderstone Report, 1982); community engineering works such as pumps, pipelines, evaporation basins (favoured by construction departments and enterprises); a system of transferable quotas for land clearing and irrigation and of transferable rights for salinity disposal (Simmons and Hall, 1990); a system of common property rights (Hodge, 1982, and Quiggin, 1986 and 1988); and regulations which restrict land clearing and irrigation practices, and which require drainage. In the context of the paucity of information about the salinity problem, each of these options face formidable implementation hurdles (see, in particular, Wills, 1987), and they have different implications for economic efficiency and for the distribution of winners and losers (see, for example, the chapters by Kirby and Blyth and Chisholm in Chisholm and Dumsday, 1987). Even so, some observations can be offered.

The technical difficulty and prohibitive cost of making direct measures of the off-site cost of each farm's land clearing and irrigation activities to salinity means pollution taxes and regulations on levels of salinity are unlikely to be practical with our current level of knowledge. If so, the focus of government intervention would have to shift from the output externality onto taxing and regulating inputs and management practices closely associated with the externalities—in particular, land clearing, irrigation and drainage activities. While the focus on inputs helps reduce monitoring problems, it comes at the cost that the relationship between input levels and off-site land degradation costs becomes further blurred—and no doubt this relationship varies widely from one region, farm, or paddock to another. Therefore, the application of taxes, subsidies and regulations on inputs which are closely related to off-site salinity (but only loosely related) cannot hope to achieve the optimal level of land degradation, although they may improve matters. For example, general taxes on irrigation (say $5 per megalitre) or subsidies on tree planting (say $100 per hectare of trees) are likely to be too little for some farms and paddocks and too much for other farms and paddocks.

For reasons of redistribution of national income, farmers clearly prefer subsidies to taxes and regulations, and in most cases they would prefer regulations to taxes. However, national efficiency concerns would place taxes at the top of the list. Off-site salinity costs are a social cost attributable to land clearing and irrigation activities. A tax on these activities directly reflects the spillover costs, albeit imperfectly. Part of these taxes and costs are passed on to buyers of farm products and so social consumption costs are more fully reflected to consumers. The deadweight costs of raising taxes to finance subsidies are avoided. Because the relative importance and magnitude of spillover effects of tree clearing and irrigation vary widely—especially regionally, but also from farm to farm—an optimal tax system on these inputs also would vary from region to region and from farm to farm.

Where regulations—for example, to restrict aggregate hectares of trees to be cleared, to require hectares of trees to be planted and to restrict aggregate usage of irrigation water—are proposed; or where a common property system is to be implemented; transferable quotas and a formal market for trading quotas should be a part of the package. The flexibility of transferable quotas leads to efficiency gains, and ready transferability may enhance distributive goals (see, for example, Simmons and Hall, 1990). Quotas would be purchased by those who value the right to irrigate and to clear trees the most from those who place a lower valuation on these rights. Since such trades are a voluntary transaction in which both the buyer and the seller gain, welfare of all in the agricultural sector as well as national welfare would be improved.

In general, the optimal tax, subsidy, or regulation will change over time as well as from place to place. As noted in the economic model discussion, the level of land degradation which goes along with the best use of the nation's scarce resources will vary with different output prices, input prices, technology, and knowledge about the technological relationships linking farm management decisions and land degradation. All these variables are in a constant state of flux in Australian agriculture. Here, constant tax, subsidy, or regulation policies will be sub-optimal.

Recognition of the complexity and difficulty in implementing government policies influencing salinity cautions against further government intervention. There is a high risk that other distortions and losses to national welfare will result, and that the cure will be more costly than the ailment. Developing a welfare improving and practical policy strategy remains an area for careful research. It will need to integrate economic guidelines and the realities of salinity as part of a careful benefit cost assessment of policy alternatives.

Concluding comments

Commercial agriculture in Australia can and does cause different types of land degradation. The most important forms are soil erosion, salinity and water logging, and soil fertility loss. These different forms of land degradation have different properties and they call for different policy responses.

In the cases of soil erosion and of soil fertility loss, in the overwhelming majority of cases most of the social costs of land degradation are on-site costs; any off-site costs are restricted to a small number of neighbourhood farms and other activities. With these two forms of land degradation, the majority of the social costs—whether they take the form of lower levels of productivity in future years or of reductions in land values—are internal to the individual farmer. Rational farmers, with access to information about the causal links between different management practices and levels of soil degradation, will choose the same mix of clearing, grazing, cropping, land degradation and land restoration activities as is desired by society if markets are allowed to perform. Government has a role in reducing distortions to product prices and input costs; reducing tax incentives and disincentives; and implementing drought policies and land tenure arrangements which will facilitate the operation of market forces to achieve socially desirable outcomes.

Decision choices of farmers and the ensuing patterns and magnitude of land degradation will change over time as commodity prices, input costs, knowledge about the links between agricultural practices and soil degradation, technology, and so forth, evolve.

The situation is more complex for land degradation taking the form of salinity and other forms where off-site costs are an important share of total land degradation costs. In the case of salinity, the off-site costs of clearing trees and of irrigation are important; the externality is of a diffuse or non-point form, and knowledge about the magnitude of externalities caused by individual farmers is limited. Rational decisions by individual farms will ignore most, if not all, of the off-site land degradation costs and result in levels of salinity which are excessive from the point of view of the best use of the nation's resources.

Significant reductions in salinity could be achieved if government were to continue eliminating the distortions of current commodity price and input cost policies. In particular, a number of irrigation intensive products receive high levels of nominal and effective rates of assistance relative to dryland optional commodities, and irrigation water is heavily subsidised. While such changes clearly would reduce land degradation, salinity still would be excessive from a national perspective.

Because of the complexity of the off-site or externality property of salinity, it is difficult to be confident about the form and magnitude which additional government intervention should take; and about whether that extra intervention would actually improve matters net of costs. Marked differences between individual farm situations mean that any general set of taxes on irrigation and land clearing, or subsidies on tree planting and drainage, or regulations on irrigation and tree clearing, are likely to be too much for some farms and too little for others. Fortunately, while this form of land degradation is serious in local areas, in the total Australian agricultural scene it affects less than 2% of the area and under 10% of total production.

5. Wood, wildlife and wilderness: managing Australia's native timber forests

Executive summary

A carefully structured move towards a more market based forest industry could achieve an improved balance of environmental and economic goals. By allowing the owners of forest resources to trade-off competing forest uses at prices which reflect real market values, including values placed on environmental amenities, Australia could have both a more efficient timber industry and more environmental benefits.

From late in the last century, State Governments in Australia have pursued a policy of public ownership and management of native forests. Political allocation of the native forest resource was largely motivated by perceptions of possible timber shortages. Today, 73% of Australia's 41.3 million hectares of native forests are owned and managed by state departments and forestry commissions. Less than 30% of the public inventory of native forests are commercially logged; however, these represent almost all the hardwood forests managed for timber production.

In recent years there has been an often acrimonious debate over the continued logging of these state owned hardwood forests. A number of the forests are listed in the National Estate by the Australian Heritage Commission. The Tantwangalo and Coolangubra state forests in south-east New South Wales, and the forests of East Gippsland are notable examples. Various conservation groups, such as the Australian Conservation Foundation (ACF) and the South-East Forest Alliance (SEFA), have advocated a total ban on logging in the National Estate areas. In pursuit of their objectives the environmental groups have engaged in vigorous protest actions, including blockading of logging operations. By focusing on all or nothing choices the debate has lost sight of the gains available from trading-off competing forest uses, including environmental and recreational amenities.

The goal of forest management should be to balance competing forest uses by factoring into decisions community valuations of timber, leisure amenities and other needs. Outcomes should not be restricted to one hundred percent timber production or one hundred percent forest preservation. Instead, each forest or stand within the forest should be put to its highest valued combination of uses. The problem is finding policies which will generate the incentives and information necessary to move towards this goal.

In this chapter, different management strategies and institutional arrangements for Australia's publicly owned native hardwood forests are evaluated against the ideal of an efficient balance between timber production and forest preservation. We focus on the forests which are currently open to commercial logging. The strategies and policies evaluated include:

- *traditional 'sustained-yield, even flow management', as commonly practised by the forests services;*
- *economic management for timber production alone;*
- *'scientific' multiple-use management with public ownership;*
- *vesting of private rights to forests in conservation groups and the timber industry, and the use of market based prices.*

It is found that, compared to sustained-yield management, economic management for timber production alone could lead to fewer forests being logged, but with the remaining forests being managed more productively. While an improvement on sustained-yield, recent moves towards multiple-use management by the state forest services also have little hope of determining community valuations accurately. As with other economic activities, it is the vesting of private ownership in the forest, and the use of market based pricing mechanisms, which generates incentives to discover the combination of uses which best suit the economic and other needs of individuals in the community.

**Chart 5.1
Native forest areas by forest type, 30 June 1986**

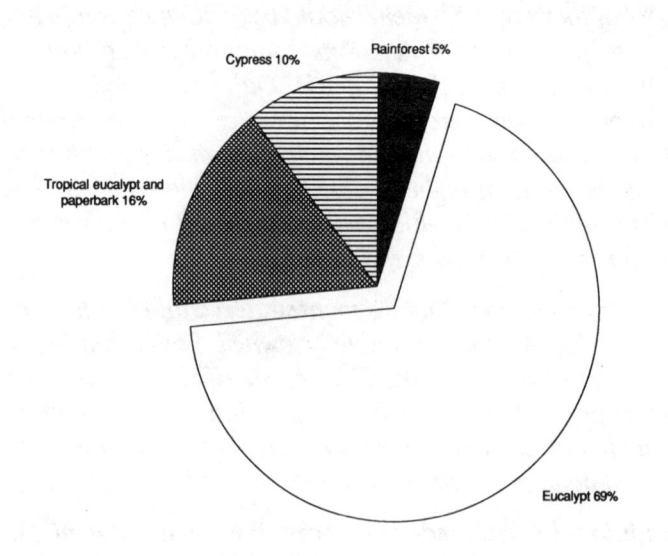

Source: 1988–89 Year Book, Australian Bureau of Statistics

The native forest resource

Native forests currently cover 5.3% of Australia's land area or 41 million hectares. (NAFI, 1989). As can be seen in Chart 5.1, the predominant forest ecosystem in Australia is the hardwood, eucalyptus forest, which makes up 28 million hectares or 68% of total native forest area. The remaining forests are classified as tropical eucalyptus paperbark, cypress pine or rainforest.

Eucalyptus forests are 'ecologically robust'. The trees are peculiarly adapted to ecological disturbances such as drought, fire and logging. Dormant buds, at the base of the tree and along its branches and stem, enable rapid regrowth when the foliage or the trunk is destroyed. Seed germination usually requires old growth to be cleared away and the creation of a nutrient rich seed-bed from fire debris. When disturbances to the forest ecosystem are managed, a range of diverse species' habitats are created with timber stands of different ages. In contrast, a forest subject to uncontrolled wild-fire tends to have uniform age classes. Some forms of wildlife, such as the Leadbeater's possum, flourish best in an environment consisting of both old growth and new growth forest. Logging practices can mimic natural disturbances to the ecosystem in a controlled fashion, and there is a strong propensity for timber harvesting in eucalyptus forests and ecological values of the forests to be mutually reinforcing. Forest management has improved the condition of some old growth forests harvested since the early 1900s. Dr Ross Florence, comments:

> ... these same forests are more dynamic, more structurally and biologically diverse and to many people more aesthetically attractive than the old-growth forests they replaced. It is a testimony to the resilience of the forests that after 100 years or more of logging, they are willingly incorporated into national parks or actively sought for inclusion on the Register of the National Estate. (Florence, 1989: 4)

Dr Tim Flannery of the Australian Museum has also expressed the view that much of the ecological damage to native eucalyptus forests in the last 200 years has occurred because Aborigines discontinued their limited 'forest management' practices of 'firestick' farming (Flannery, 1989).

Unfortunately our perceptions of timber harvesting are shaped by images from the past, which are often distorted. Since European settlement in the 19th century, 50% of native forests have been cleared. From our vantage point, with native forests relatively scarce and wildlife and wilderness values increasingly important, the early clearing of native forests appears wanton. Much of the blame is apportioned to an unfettered, 'free market', timber industry; but such a conclusion is unwarranted for the following reasons.

1) The major portion of timber felling was to clear land for agriculture. At the time, agricultural land was scarce relative to wood, wildlife, and wilderness, which were in abundant supply. Clearing the land

was often a rational management choice given relative valuations of the alternative uses.

2) Property rights to forest resources were inadequately defined. Consequently, some valuable forest resources were destroyed. In chapter three, the examples of red cedar in the Hunter Valley and the Huon pine in Tasmania were mentioned amongst others. However, rights to these timber resources were either vested entirely in the crown—which was powerless to police them—or else, valuable privately owned timber stands were cut out by timber gangs before the settler physically occupied his land.

3) The early settlers had little information about an unfamiliar environment on which to base management decisions. Problems such as salination, destruction of biotic diversity, and the importance of forests for combating greenhouse effects never entered the decision-making process. Here, governments were as ignorant as markets, as evidenced by the subsidies and taxation incentives later provided for land clearing.

The appropriate policy response to a perceived failure in the timber market should have been a more adequate definition and monitoring of private rights to forest resources. Then, as timber became more scarce, an increase in the price would have created the incentive to expand long run timber supply. As we argue later, private management of the 'ecologically robust' eucalypt forests, would have provided a more productive and less wasteful timber harvesting regime. A case could have been made for government subsidising research to increase knowledge of the forest ecosystems, and for the setting aside of National Parks for preserving particularly outstanding examples of natural beauty; however, the Australian state governments went much further than this.

Anticipating a long run timber shortage and ignoring the role of price, the various state governments in Australia followed the

Chart 5.2
Native forest areas classified by ownership, 30 June 1986

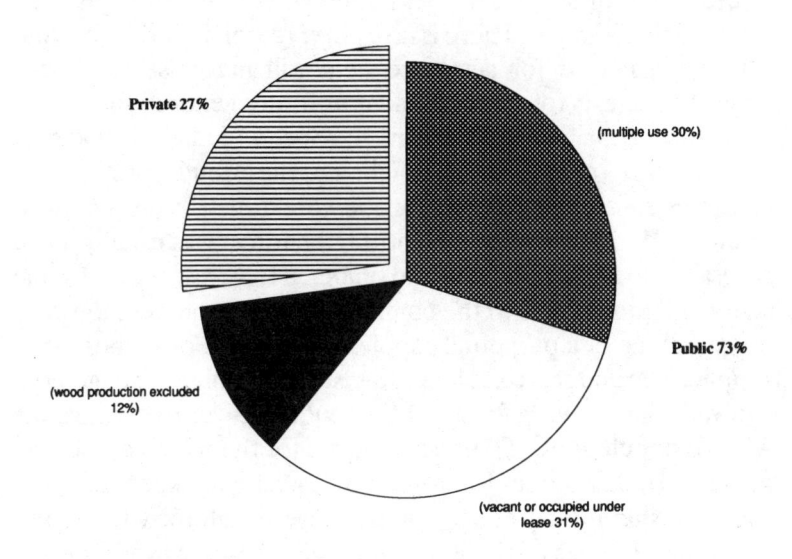

Private 27%
(multiple use 30%)
Public 73%
(wood production excluded 12%)
(vacant or occupied under lease 31%)

Source: 1988–89 Year Book, Australian Bureau of Statistics

European and US pattern and set aside state forests for commercial timber production. Forest services were set up to manage the resource, and a strong precedent for state management of forest resources was established (see chapter two for a more detailed discussion). Today 73% of the native forests are publicly owned (see Chart 5.2). Less than 30% are managed for timber production, but these represent almost all of Australia's commercial hardwood forests.

Traditional sustained-yield, even flow management

Public management of the Australian native forests followed the traditional European sustained-yield, even flow model. The principal of sustained-yield has to be clearly distinguished from the more contemporary term 'sustainable development'. Sustained-yield is a structured timber harvesting programme which is designed to yield the maximum annual volume of wood from a forest in perpetuity. It takes no account of price, alternative uses of the forest, or the interest and other costs of investing in fixed assets such as timber. There is no *a priori* reason for believing that a forest managed for sustained-yield will maximise the social benefit of forests for either present or future generations.

The concept finds its roots in European feudalism. Dowdle (1981) gives us an interesting insight into these feudal origins. For centuries, forest resources in Europe were owned by the local lord or duke. While timber was plentiful, the forest was usually made accessible to all for firewood and timber. However, without some system limiting access to the common timber resource, there was no guarantee of a perpetual supply of fire-wood, or timber, once supplies became scarce. Hence, a system of timber management evolved which closely followed the feudal answer to diminishing wildlife populations. To preserve game for the King's chase, bag limits were defined and enforced by forest-gamekeepers. In a similar fashion, 'bag limits' for trees were enforced to ensure perpetual supplies of timber. Eventually these evolved into the sustained-yield, even-flow management strategy which early Australian and North American foresters imported from Europe.

Under sustained-yield, even flow, the land base for timber harvesting is taken as given. No economic criteria are used to evaluate the forestry returns from land with different levels of fertility and accessibility. Forests with negative net returns are harvested alongside forests with positive net returns. Decisions relating to the extent of the land base are inherently political.

Once the forested area available for harvesting is determined, the goal is to create a regulated forest which yields the highest possible, constant volume of timber each year. First, the timber harvest age is determined by calculating the age at which the mean annual increment to timber volume peaks. Chart 5.3 graphically illustrates the correct harvest under sustained-yield management. We assume that timber follows a typical, S-shaped growth curve with timber volume growing slowly at first, in the young forest, then rising rapidly before tailing off as the forest reaches maturity. The slope of a straight line extending from the origin to a point

Chart 5.3
Sustained yield harvest age

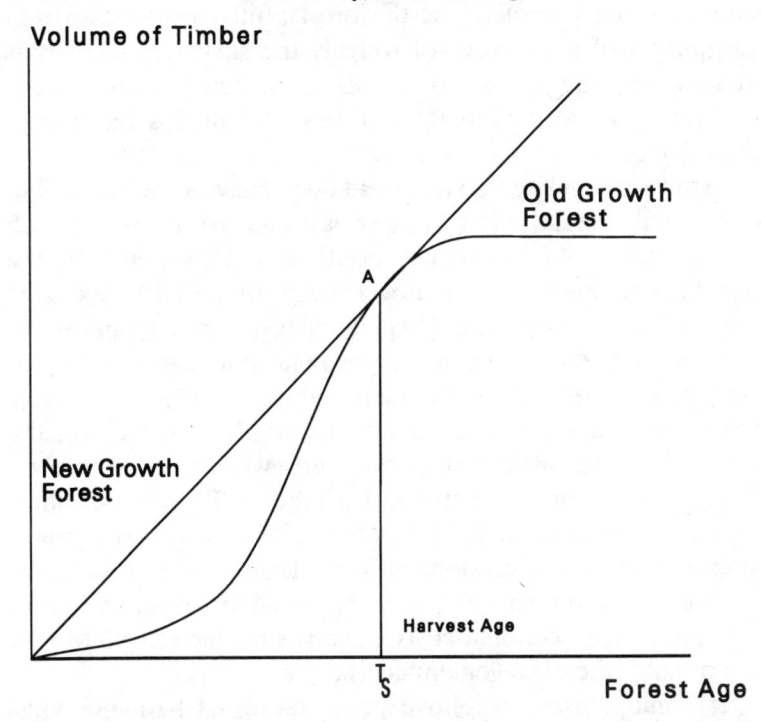

on the yield curve measures the average rate of growth or mean annual increment. When this line is steepest, at point A on the yield curve, mean annual increment is at its peak and the required harvest age is T_s. Maximising mean annual increment for each growing cycle ensures that, on average, the highest possible volume of timber is harvested each year; but, as we explain in the next section, only when real interest rates are zero will it correspond to a rational, economic, harvesting decision.

Having maximised sustained-yield, all that remains is to ensure that the timber comes in an even flow. To achieve an even flow of timber, foresters work towards creating a regulated forest which has a relatively even distribution of timber stands in every age class. If we assume equal productivity over the whole forest, dividing total forest area by harvest age, T_s, gives the forest area required for each age class. It also tells us the area of forest that should be harvested annually for an even flow of timber volume. Variations in productivity may require some adjustments to the harvested area; however, once the forest is fully regulated, an even distribution of age classes of roughly the same productivity is created, and the volume of timber harvested is equal to the maximum annual growth rate. Foresters call this the annual allowable cut.

In the 'un-regulated' state, forests do not have an even distribution of timber stands of every age class. Old growth forests, such as those in East Gippsland or South East New South Wales, include many stands of timber over the maximum sustained-yield harvest age, T_s. To create a 'regulated' forest, the annual allowable cut it set equal to the *current* average annual growth rate for the forest a whole. Timber is harvested from the slower growing excess inventories of over-mature timber first; therefore, the average rate of growth for the whole forest will increase over time. Eventually, the forest will approach a 'regulated' state, and annual growth will be maximised. However, the harvesting of old growth forest is not linked to economic criteria. It only serves to maximise the volume of timber cut under the sustained-yield, even flow constraint. The economic costs, of harvesting old growth forests, do not enter the management decision.

In some senses the above presentation of sustained-yield management is too stylised. In practice, the theoretical maximum wood volumes generated by the sustained-yield model are reduced to take account of nature reserves, habitat trees for wild life, limits to logging near river banks and other general provisions of The Forest Practices Code. However, these reductions, can tend to reflect bureaucratic rather than economic judgements.

Sustained-yield remains a guiding principle for the various state government departments which manage Australia's native forests. Professor Ian Ferguson, who conducted the Board of Inquiry into the Timber Industry in Victoria, called it a sensible management tool in the in the 19th century, but quite inappropriate now.

> Sustained-yield embodies a physical and static objective, being based on the maximisation of average wood produc-

tion at a constant rate in perpetuity. Wood however is an economic good with many substitutes, there are many alternative uses for these resources and not enough of them. It cannot be assumed that more wood via maximisation is desirable or that a constant amount is sensible. There is ample evidence that a constant amount is an unworkable or at least an inefficient solution, because markets are in a perpetual state of flux rather than a steady state. (Ferguson, 1987: 35)

Economic models of timber production

Economic modelling of the timber production decision has a long and rich history. As far back as 1849, Faustmann formulated the basic equations for calculating the optimal economic harvest age. He assumed a single stand of timber, managed in perpetuity, with known and unchanging prices, interest rates, costs, and timber yield curves. The Faustmann model, as it came to be known, was the building block for a range of increasingly complex and diverse models, which considered other aspects of the management decision, such as optimum thinning age, and multiple-use management for recreational and amenity value as well as timber production. The mathematics of these models are beyond the scope of this chapter (For those interested in a more in depth presentation, Samuelson (1976) gives a detailed discussion of the Faustmann model, Nguyen (1979) discusses an application of the model which incorporates optimum thinning ages, Hartman (1976) generalises the model to include non-timber services on a single timber stand, and Bowes and Krutilla (1985) develop a multiple-use management model which considers a number of timber stands with varying age classes).

The simplest case of efficient economic management considers timber production alone. Both the theoretical results and some empirical evidence from the US indicate that, compared to economic efficiency, managing for sustained-yield leads to:

a) a longer growing period with older and slower growing timber stands;

b) more harvesting of old growth forests than is justifiable economically;

c) and over investment in new growth forests.

The timber harvesting decision involves maximising the net present value of an asset which is going to be sold at some time in the future. For a single harvest cycle, consider the S-shaped growth curve shown in Chart 5.3. As a stand of timber of timber approaches maturity, its rate of growth begins to taper off. If payment is made in proportion to biomass or volume of timber, a point will be reached where harvesting the timber and investing the money at market interest rates, will give a higher return than waiting a year and selling a slightly higher volume of timber. The sensible thing to do is harvest the timber when the rate of increase in revenue from selling an increased timber volume falls to the market rate of interest.

Extending the analysis to multiple harvest cycles introduces an added complexity. Delaying harvest by one period now has two opportunity costs. Harvesting timber a period earlier not only allows money to be placed in the bank, but also enables an earlier planting for subsequent timber harvests. For the complete picture, timber should be harvested when the value of an additional increase in timber volume, from the next period's growth, falls to the level of the interest income obtained from banking the proceeds of timber harvested in the present period plus the interest benefit of bringing forward all subsequent harvest cycles. The Faustmann model is a mathematical solution to this problem.

Comparing the sustained-yield even flow harvest age with the Faustmann harvest age, it can be shown that the latter Faustmann age almost always offers a better return. The one exception is when interest rates are zero. In this case returns the will be equal. If timber production is economically viable and interest rates are positive, harvesting for economic efficiency will usually lead to younger stands of timber. However, when costs outweigh benefits at every harvest age, timber harvesting will not be carried out at all. Economic management for timber production alone could lead to abandoning timber production in some areas of the forest, while increasing returns in more productive areas through shorter harvest ages. In contrast, sustained-yield management will always lead to timber harvesting, because prices and the opportunity costs of investing the returns are not considered.

The sustained-yield even flow model can also lead to incorrect evaluations of forest investments such as planting seedlings, thinning and fertilising. Dowdle (1981) discusses how the model makes returns from one part of the forest justify investments elsewhere in the forest. Consequently, forest managers are led to accept investments more conventional financial models would reject.

Public timber management—a record of economic inefficiency

Managing for timber production alone could lead to greater forest preservation in some areas, and more intensive and efficient production in other forest areas. In the absence of a market for wildlife and wilderness, the efficient economic solution for timber production approximates the free-market solution. The difficulties and implications of managing forests for multiple-use later will be discussed later. First we examine some of the evidence which supports the claim that a free market for timber could result in a more productive timber industry and better environmental management than under traditional systems of political allocation.

There is some compelling evidence from the US that political allocation of forest resources has created an inefficient forest industry and over-emphasised timber production. The Australian evidence is more circumstantial and somewhat less rigorous, and it seems clear that studies similar to those in the US need to be conducted. These would be a valuable contribution to the debate over Australia's more controversial forest management issues—for example, the East Gippsland National Estate Forests.

Public forest management in the United States

The US Forest Service is the largest natural resource agency in the federal government, with an annual budget of around $2 billion and roughly 39,000 full-time employees. It oversees natural resource use on 191 million acres of national forests; and is required by law to manage its lands for multiple-use—which includes timber production, livestock grazing, mineral and energy production, fish and wildlife production, wilderness protection, and public recreation. In the past two decades the agency has come under a barrage of criticism both from environmentalists (see the National Audubon Society, 1986), who feel that the agency has over-emphasised commodity production at the expense of environmental amenities, and from commodity based interests who feel that wilderness values are receiving too much attention.

Compounding environmental concerns are fiscal concerns; virtually every aspect of national forest management loses money. For example, the seven national forests surrounding Yellowstone National Park had losses from their timber programs ranging from $241,000 per year to $2.2 million per year from 1979 through 1984. (The Wilderness Society, 1987: 22). Less well recognised are the losses on recreational services. The millions of people who fish, hike, and hunt on national forests generally do so at zero cost to themselves. On the Gallatin National Forest alone expenditures surpassed receipts by nearly $2 million, approximately twice the deficit from timber production, and this is typical throughout the public forests. (See Anderson et al., 1990: 23).

Until recently forest service management decisions were primarily based on the sustained-yield even flow model outlined previously. Analysis of forest service management decisions, using economic models of timber production, has shown that forest service agencies have been harvesting too much timber in previously unharvested forests and inefficiently managing timber production in existing commercial timber forests. Hyde (1981) examined forest service timber harvesting from the San Juan National Forest in Colorado—a forest which covers 1.85 million acres. First, Hyde considered the economic viability of a forest service plan to bring approximately 400 thousand acres of marginal old growth forest into timber production. He concluded that the timber price necessary to bring this land into production would need to be at least $45.45 per board feet in 1976 dollars. Average sale prices in that year were only $2.65 and the highest average over the previous was only $23.15.

Next, Hyde used the Faustmann model of timber production to analyse perpetual timber harvesting on currently managed forests. He calculated the price that would be necessary for the forest service to achieve a positive return from perpetual harvesting of the forest lands in question. Some adjustments were also made for compatible uses of the forest. Even when these were taken into account, the required price was $26 per board feet at a 7% discount rate. As average annual sale prices had not risen above $23.15 over the previous six years, only highly valued timber sales were justifiable. Hyde concluded that:

> ... under current practice, more land is used than can be justified by a free timber market. Environmental destruction and foregone recreational opportunity result. (Hyde, 1981: 200)

State Forest Service management in Australia

In aggregate, Australian state forest management agencies have been making substantial operating losses for the last 50 years. Bruce (1985) estimated that average annual receipts amounted to 61% of expenditure or less from June 1966 to June 1982. These aggregate estimates did not separate capital from current expenditure, nor expenditure on commercial timber production from that on other non-wood values. However, Walker (1984), went a step further and attempted to assign expenditures to their appropriate categories using Tasmanian Forest Service Budgets. His analysis indicated that, in the four years to 30th of June 1982, taxpayers' contributed 16% of total receipts of the Commission—which was twice the estimated cost of recreation and wildlife conservation services. When discussing the provision of market goods such as timber from Victorian State Forests, Ferguson comments:

> To date, these goods have been supplied with little reference to the economic viability of the use of those resources. Indeed, the previous accounting/budgeting system of the Department of Conservation, Forests and Land did not even enable such matters to be evaluated properly (Ferguson, 1987: 37).

In recent years there is some indication, at least in some states, that the operating losses of forest services are being turned around. In New South Wales, however this maybe illusionary. The Forestry Commission 1987–88 income was $63 million and its expenditure $50 million, according to the Public Account Committee (1990) this may due to subsidies or more importantly our effective interest forgiveness of $14 million. The Committee also gathered persuasive evidence of considerable cost padding in the New South Wales Forestry Commission.

Traditionally, forest service agencies have followed sustained-yield management in most state forests managed for timber production. The theoretical analysis of sustained—yield management in the previous section, the large budgetary deficits of the state forest services, and the analogous situation in the US, all point towards wasteful investment in timber production from Australian native forests. Two empirical studies support this view.

One study, used by the Australian Conservation Foundation (ACF) in its advocacy of discontinuing logging in native forests, indicated that investing in a representative 100 hectares of native forest yields a negative net present values at any discount rate above 4% (Cameron and Penna, 1989). Chart 5.4 summarises net present values of native forest investment for a range of discount rates. The ACF used this data to argue for an almost complete ban on logging in native forests. There are at least two pit-falls in such a conclusion. Firstly, the data are averages and represent a summary of a range of investments with variable net present values. While the average indicates that new investment should be scaled back, there are still likely to be some investments with positive net present values at closer to market rates of return. Secondly, the new investment calculations do not consider the value of harvesting the standing stock of timber. In Chart 5.4, this value is shown by the line A–B. It is positive at any discount rate. In contrast to Hyde's conclusion for the San Juan forest in Colorado, it appears economically viable to harvest at least some of the standing stock of timber from Australia's native eucalyptus forests. However, the figure is, once again, an average intended to be representative of all native forests in Australia. Logic dictates that there will be some productive and accessible native

Chart 5.4
NPV of new investment in native forest at different discount rates and the value of current native timber stocks (100 representative hectares)

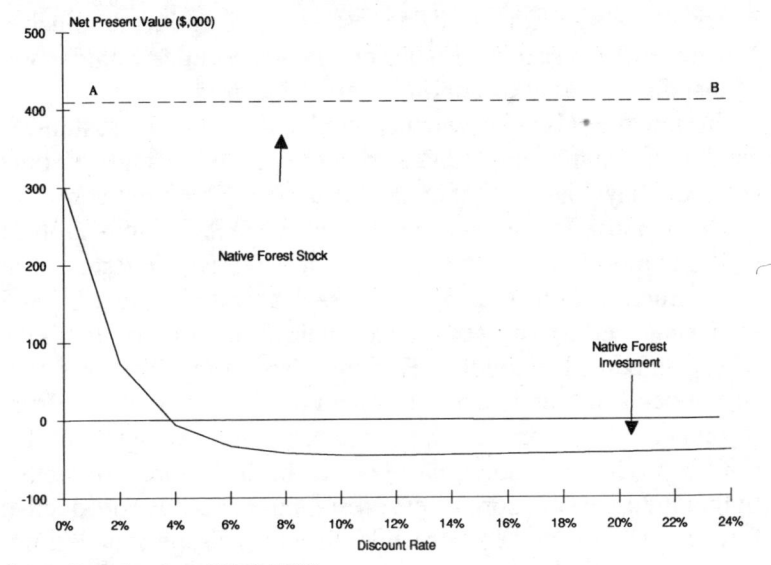

Source: Base data from INFM (1987)

forests in Australia which currently have economically harvestable standing stocks of timber; but there will also be some marginal forests which a careful economic analysis could reject for timber harvesting.

Another interesting view on Australian Native Forest timber management is given in the IC's recent report on Recycling (IC, 1990: 141–156). An appendix to this paper calculated the optimal harvest age for a range of eucalyptus forest types in various locations. A variation of the Faustmann model was used which allows for thinning of the timber stand at various ages. The unambiguous conclusion from the analysis was that the optimal harvest age, for all forests analysed, was considerably less than those adopted by the forest services concerned. For example, Ash forests in Victoria are currently harvested at around 80 years, while the optimal harvest age from a wood production perspective is 20 years. At this age timber would only be harvested for wood pulp. To ensure a supply of sawlog timber, prices for sawlogs would need to be higher or more closely related to timber volume. The study indicates that there is considerable potential for increasing the economic viability of timber production from some Australian native forests if harvest ages are reduced. Such a conclusion is consistent with our previous theoretical comparison of sustained-yield and economic harvesting ages.

In sum, the IC evidence conclusively shows that the sustained-yield timber management has lead to over-mature stands of some commercially logged timber in Australia. The representative model of native forest investment cited by the ACF lends support to the contention that there is considerable over-investment in native eucalyptus forests. Practices such as thinning, seeding and fertilising need to assessed on economic criteria not on the basis of sustained-yield models. Evidence from the US shows that sustained-yield management can lead to harvesting timber from too much of the forest. Therefore economic management, for timber production alone, can lead to both a more productive timber industry and more forest preservation, because productive forests are harvested earlier while logging is discontinued in unproductive forests. Currently, however there are no studies which have tested this suggestion rigorously for marginal forests in Australia.

Managing for multiple-use

In recent years some forest service agencies have moved away from sustained-yield management plans towards more 'scientific' multiple-use planning procedures which incorporate the methodologies of theoretical economic models. However these models are calibrated and estimated in an informational vacuum. Without information about peoples valuations of environmental amenities, solving for optimal outcomes becomes extremely difficult. The only reliable way we know of eliciting information about people's preferences is to charge a price. Currently, environmental amenities are supplied at minimal or zero price, but rational allocation based on such prices is an impossibility.

The simplest form of a multiple-use model is that proposed by Hartman (1976). This model forms the basis of the US Forest

Service, harvest-scheduling model, FORPLAN. Use of FORPLAN and related multiple-use planning procedures have been advocated in Australia (see Turner, 1987). Attempts have been made to apply the approach in forests such as the Coolangubra State Forest, east of Bombala, New South Wales.

The Hartman model is very similar to the Faustmann model for timber production. It solves for the optimal harvest age on a single stand of timber but adds a flow of non-timber values which is related to the age of the stock of timber. The solution requires that any change in environmental benefits be added to the value of increased timber volume from delaying harvest one period. Optimal harvest age comes when the combined value of increased timber volume and any change in non-timber benefits just equals the costs of delaying current and future harvests. [Readers are referred to Bowes and Krutilla (1985) for a more complete discussion of the effects of differing rates of increase in environmental benefit on the model's solution]. In brief, when recreational and amenity values increase consistently over time, the Hartman harvest age will exceed the Faustmann harvest age. However, not all amenity values increase with harvest age, implying that a later harvest age need not always be the case.

In the 'ecologically robust' eucalyptus forest, some environmental values decrease over time. As mentioned earlier, a mature forest will have a lower level of biological and structural diversity.

Carbon dioxide fixation adds a further complication. Young growth forests fix more carbon dioxide (through photosynthesis) than old growth forests. Carbon dioxide is considered one of the principal greenhouse gases (see Chapter 10). Trees play a major role in removing the gas from the atmosphere through photosynthesis. The amount of carbon dioxide removed (or, more correctly, fixed in to cellulose) is directly related to the growth rate of the forest. Chart 5.5 shows the growth rate for a typical eucalyptus timber stand. Florence (1989) states:

> If fixation of atmospheric carbon were an important objective of forest management, it would best be achieved by clear-felling, using mechanical site disturbance to create a seed-bed, thinning regularly to maintain large tree crowns

Chart 5.5
Growth pattern for Eucalyptus globulus on a high quality site

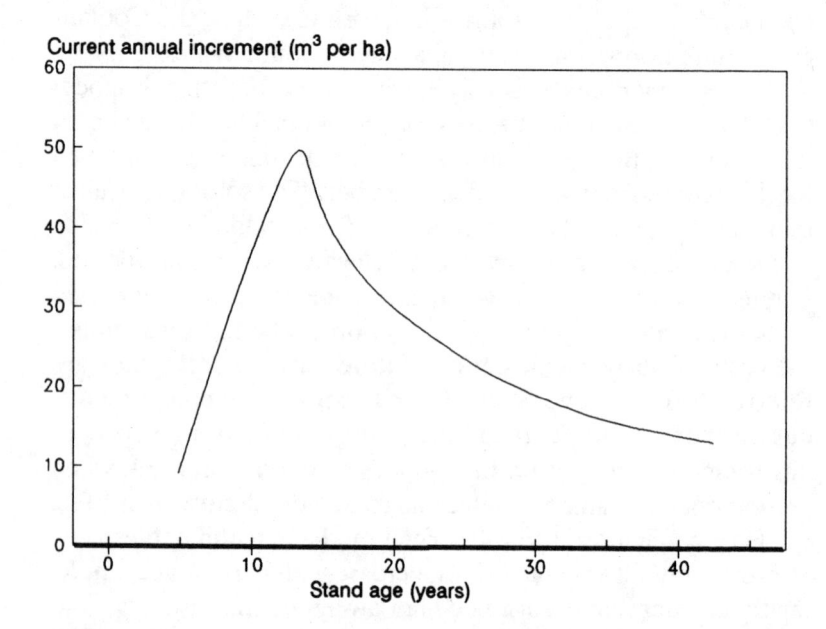

Current annual increment (m³ per ha) vs Stand age (years)

This 'current annual volume increment' curve also serves to indicate the rate of fixation of atmospheric carbon dioxide. All fast-growing eucalypt species have growth patterns of this nature — though for some, peak production will be earlier (e.g. *E. tereticomis, E. grandis*) and others somewhat later (e.g. *E. regnans, E. pilularis*). Slower growing eucalypts — including those better adapted to infertile or dry sites — may not have a sharp production peak (e.g. *E. obliqua*) — though maximum volume increment (or carbon dioxide fixation rate) will still occur by age 30 to 40 years.

Source: Florence (1989: 13–14) (Data from Goodwin and Candy, 1986)

and vigorous growth and harvesting on a relatively short rotation.(Florence, 1989: 3-14)

Cutting a forest does decrease the store of carbon locked up in old growth forests. However, unless the timber is burnt, it moves from a natural carbon 'bank' to a man-made carbon 'bank' in areas such as house and furniture construction. Arguably, with the aid of preservatives and other decay inhibitors, this man-made carbon store could hold carbon longer than an old growth forest. The contribution forests can make for fixation of greenhouse gases and is, however, easily exaggerated. Beckerman (1990: 21) estimates that even if the whole world's tropical forests were to be immediately burnt, the concentrate of CO_2 would rise by only 5-10%.

Forage for wildlife, prevention of wildfire, and increased water flow are other amenity values that may decrease with stand age (see NAFI, 1989: 8–16). Obviously some amenity values, such as wilderness or the value individuals place on undisturbed ecosystems, will increase. However, no *prima facie* case can be made that harvesting on a sustained-yield harvest age either increases or decreases environmental quality.

Bowes and Krutilla (1985) further refine the Hartman model and develop a theoretical model of multiple-use which allows harvesting or preservation decisions on one timber stand to affect timber and amenity values on other forest stands. The added level of complexity only reinforces the conclusion that sustained-yield, even-flow management does not optimise either timber production or amenity values, even when amenity values are taken into account.

Long even flow rotation, far from being the desirable compromise policy for multiple-use management, may simply provide both uneconomic timber and a poor balance of age classes for non-timber use. Perhaps most importantly we see ... that the harvesting decision can be extremely sensitive to factors about which we have little empirical knowledge. (Bowes and Krutilla, 1985: 567)

Unfortunately, scientific multiple-use management by political agents is no alternative to sustained-yield management. The theoretical models of multiple-use point to a problem without proposing a practical solution. Without prices and markets for amenity values from the forests, we cannot know what the optimal rotation or harvesting decision is. How can we know how much individuals truly value recreational opportunities and wilderness values, other than by observing how much they are willing to pay for those activities? Merely asking them is unlikely to produce an accurate answer given that they are not required to pay the cost of their stated preference. For multiple-use management to be effective, the role of markets must be extended to encompass

Markets for wood, wildlife and wilderness

When examining the prospect of diminishing wilderness value in National parks because of open access to all, Garret Hardin asked the question:

> What shall we do? We have several options. We might sell them off to private property. We might keep them as public property but allocate the right to enter them. The allocation might be on the basis of wealth, by the use of an auction system. It might be on the basis of merit, as defined by some agreed-upon standards. It might be by lottery. Or it might be on a first-come, first served basis, administered to long queues. These, I think are all objectionable. But we must choose...(Hardin, 1968: 21)

We face a choice in regard to Australia's native timber forests. At present use of the resource is administered politically. The past record of public sector management in these forests is not one of efficient management for either timber or amenity values. The question is could markets, private ownership, and an allocation system based on price do better?

Charging a price for wood does not go against anyone's sense of propriety, but charging a price for wildlife or wilderness tends to sit uneasily in our collective Australian consciousness. A nation, cushioned by years of open access to beaches, forests and the bush, reacts somewhat uneasily to suggestions of charges, private ownership and fences.

Nevertheless, as population growth and a more affluent society generate higher demands for scarce recreational and amenity resources in the forest, the role of market mechanisms (such as ownership vesting and market-based prices) needs to be carefully considered.

A common argument for leaving amenity values unpriced is that people who enjoy these values cannot be excluded from their enjoyment even if they do not pay. It is argued that it is much easier to exclude the logger who does not pay for the tree than it is to exclude the hiker who does not possess a trail pass. But finding ways of charging a price to those who enjoy benefits from assets is precisely the role of the entrepreneur.

Entrepreneurial imagination is crucial to a market approach. Where environmental entrepreneurs can devise ways of marketing environmental values, market incentives can have dramatic results. What is important is that any case of external benefits or costs provides fertile ground for an entrepreneur who can define and enforce property rights. A timber owner who can devise ways of charging bush-walkers benefiting from the resource can internalise the benefits and costs. By receiving payment from those who enjoy the bush-walking, the new owner has an incentive to maintain or improve resource quality. A conservation group which can enter the bidding process for the rights to a transferable forest resource can capture the preservation value of the resource and utilise limited timber production, compatible with non-timber uses, to finance other bids. Creative marketing of forest resources and wilderness values, through sale of hunting, bush-walking and observation rights, would again internalise costs and benefits and overcome supposed free-rider problems.

It is not even necessary that all goods provided by a forest be marketable. Often the ability to market associated goods or services will enable the entrepreneur to capture adequate benefits from providing what appears on the surface to be a non-marketable commodity. For example, when visiting a famous art museum patrons are often happy to pay a premium for lunch, because they believe 'it's in a good cause.' In effect, the museum can charge indirectly for having great works of art on public display. Similarly, a hiker on a forest trail may be willing to pay a high price—relative to the costs of providing the trail, keeping it clean and so on—because he knows his payment will contribute to the preservation of the forest, its wildlife and genetic resources. If many bush walkers are willing to make such payments, more forests will be supplied by entrepreneurs than would be the case if hiking opportunities alone were in demand.

The idea that entrepreneurs would exclude potential consumers when such consumers could be allowed to consume at zero cost also needs some scrutiny. So long as there are potential consumers who could be served at a cost less than the price they are willing to pay, entrepreneurs have an incentive to make further deals. In particular, an almost endless number of price discrimination mechanisms and non-linear pricing schedules (such as volume discounts, annual fees or multiple-site access fees) can be devised to ensure that all who wish to consume are allowed to do so at a price which compensates the owner for losses from consumers previously paying higher prices. Once property rights have been defined, and can be defended and transferred, the owners have a great and personal interest in doing all they can to help all interested parties obtain benefit from the resources under their stewardship.

Existence values

Values associated with existence of a resource are the most difficult to capture using market mechanisms. By existence value we mean the value individuals place on knowing that somewhere there are areas of forest reserved from logging. It is a value which implies vicarious enjoyment of forest preservation even if the forest is never visited. Here, one person's enjoyment cannot exclude another's; and, theoretically, market mechanisms cannot always provide a social optimal allocation. While we recognise the problem, there are a number of factors which minimise its importance. First, if there are wide differences in the strength of peoples preferences for forest conservation then many of the externalities will be irrelevant. Chart 5.6 demonstrates this for a simple case of two people with different valuations of forest preservation. Person A sets a high value on forest preservation. Line AB represents his demand for preservation. Person B sets a much lower value on forest preservation. His demand curve is represented by CD. In contrast to an apple or orange, which can only be enjoyed by one person at a time, each extra unit of forest preservation gives value to both person A and person B. In the terminology explained in chapter three, forest preservation (for existence value alone), is a non-rival or joint good. To determine the socially optimal level of forest preservation, the individual demand curves should be summed vertically (Samuelson, 1954) as the total benefit to society of each additional forest preserved is

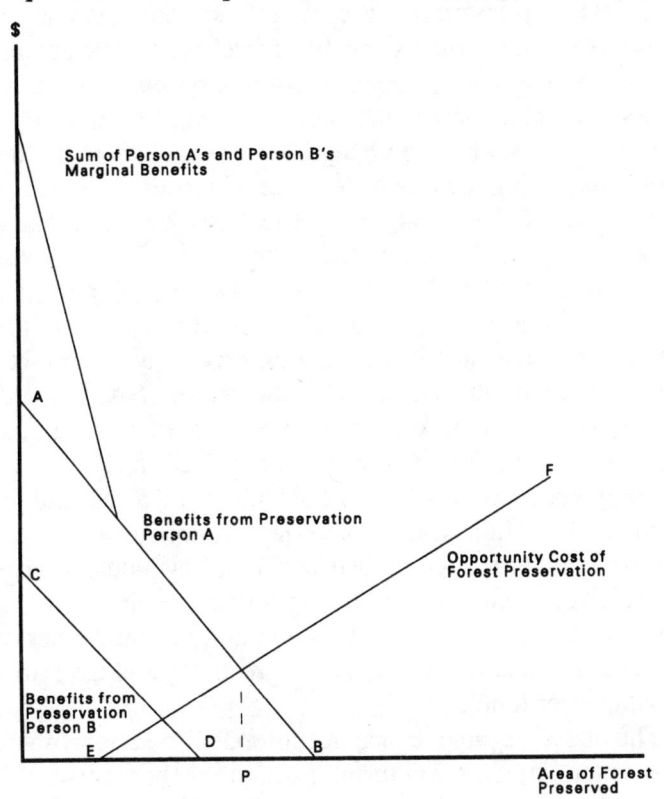

Chart 5.6
Equilibrium level of preservation with differing valuations

P = Opportunity Cost of Forest Preservation

the sum of the benefit to both person A and person B.

Line EF shows the opportunity cost of forest preservation. Curtailed timber production, with its associated benefits, becomes a lost opportunity when a forest is preserved. This cost to society should be balanced against any benefits from forest preservation. The costs increase as forest preservation increases, because more valuable timber stands are put aside for conservation reserves or national parks. The socially optimal level of forest preservation is point P, where the increased benefit from setting aside another forest is exactly equal to the social cost of preservation. The key point to note is that, because person B's valuation of the forest resource is markedly different from person A's, person B's valuation has no effect on the final optimal outcome. In a free market, even if B free-rides on A's forest purchases, the outcome will still be socially optimal; the externality is irrelevant. The presence of 'free-riders' on other peoples' consumption purchases does not always lead to socially undesirable outcomes, particularly when preferences differ by large magnitudes, as is likely to be the case when considering the existence value of a forest.

The second factor affecting the free-rider problem is the presence of other counter balancing externalities. Existence values associated with forest preservation are not the only relevant externalities that need to be considered. Timber production also has associated externalities, especially when factors of production cannot be employed elsewhere. NAFI (1990) has estimated that a 10% reduction in exports of sawmill products would lead to a $27 million reduction in GDP in 1987–88. Only part of this reduction in GDP would impact on forest and sawmill product related industries. For example, although 430 jobs would be lost in timber related industries, 270 jobs would also be lost in the wider economy through linkages with the timber and forest product sector. (However, these job and income 'losses' would not amount to sustained 'losses' as resources adjust into other activities over time).

The above arguments are not intended to support increased government intervention to subsidise the timber industry! In the first place the magnitude of the relevant externalities is not known. Attempts by central planners to regulate, subsidise and scientifically optimise are fraught with difficulty. However, one cannot argue that existence values create an externality and then not consider externalities on the timber production side of the equation. One counteracts the other and the best approach may be to ignore both rather than bear the additional costs of government intervention.

Recent NSW experience

As an example of how a market based system using private ownership might work, consider the recent decisions by the New South Wales government in relation to the South-East NSW forests.

In late 1990 the New South Wales and federal governments agreed to turn 60% of the disputed National Estate forests on the NSW coast into national parks while leaving the remaining 40% open to timber harvesting. The decision was projected to result in the loss of 11.5% of timber supply from the area, 112 jobs and one sawmill. 56,000 ha were turned into National Parks. Neither the timber industry nor the conservation groups appeared happy with the decision. A timber industry spokesman said 'the plan was a compromise which favoured conservationists'. Conservation groups such as the Australian Conservation Foundation and the Wilderness Society were also concerned, describing the agreement to 'gut' the Coolangubra wilderness as 'contrary to community opinion and ecologically irresponsible' (*The Australian*, 9th Oct, 1990).

Given the decision to withdraw this quantity of land from wood production, a market approach to the problem could have been to vest the land in recognised conservation groups. They would have had full rights to the parks with responsibility for their on going management. The remaining timber forests could have been vested in the timber industry groups which previously had logging rights in the forest. Once the initial allocation was made, forest owners could have traded their assets to achieve a more mutually beneficial outcome. If conservation groups felt the preservation value of Coolangubra was much higher than other forests they owned, agreements with the timber industry could have been negotiated to allow logging elsewhere in return for preservation of Coolangubra. Alternatively, funds could have been raised to purchase Coolangubra outright. Where restricted logging was compatible with environmental values—for example in the case of preventative fire management—conservation groups would have had an incentive to allow logging in these areas, but under more stringent environmental guide-lines if desired. Funds raised could have been used for the purchase of other conservation areas. Timber industry groups would have had an incentive to consider a wider range of values in their decision making; prices charged for environmental amenities would no longer have been under-cut locally by free government provision.

As long as rights were vested on a long-term basis, both the timber industry and the conservationists would have benefited from increased resource security. Rights could have been allocated as long term leases or perpetual but tradeable utilisation rights. The decisions on forest management would no longer have been dictated by the three year time horizon of political parties; but, instead, by the longer investment profiles of the timber companies and speculators; amongst the latter would be conservationists concerned for future generations.

In this context it is important to note that National Parks, World Heritage areas and State reserves presently reserve 70% of native forests from timber getting whilst the timber industry has only tenuous security in the less than 30% of native forests open to commercial logging.

Private provision of amenity values

In the remainder of this chapter, we put some practical flesh on the theoretical bones, and show how entrepreneurial imagination can, and has, generated markets for the amenity values of forests. As mentioned above, groups from both sides of the forest conservation debate could benefit from such a market approach. Some of the examples cited are not directly forest related, but they demonstrate the ability of private markets to provide wildlife and wilderness values. Such market propensities could equally well be generated in a forest resource context.

TIMBER INDUSTRY PERSPECTIVE

The International Paper (IP) Company's wildlife program in the south of the US is a prime example of the how a firm can benefit from using forest resources to provide non-wood services. IP employs specialists to oversee wildlife and recreation on its lands, including the 16,000–acre Southlands Experiment Forest located near Bainbridge, Georgia. There, research is carried out to develop forest management practices that enhance wildlife population as well as profits. White-tailed deer, turkeys, rabbits, bobwhite quail, mourning doves and other species are beginning to reap the benefits of new management techniques. Habitat is improved by controlled burning, buffer zones along streams, and tree cutting practices that leave wildlife cover, and plenty of forage (The Council on Environmental Quality, 1984: 426).

According to company officials, investing in wildlife research and habitat production makes sound business sense. On its 1.65 million acres in the mid-South region (Texas, Louisiana, and Arkansas), IP in 1988 charged an average of 83 cents per acre for hunting clubs and 62 cents per acre for individual hunters. Company officials see a good possibility that the return could as high as $10 per acre in the future as more hunters seek better hunting conditions available on IP lands. (Blood and Baden, 1984: 11)

North Main Woods, Inc. offers another interesting example of industry groups responding to recreational demands. A non-profit association formed by 20 landowners, North Maine Woods Inc. manages recreation on 2.8 million acres of mostly private commercial forests. The area includes two of the most wild rivers in New England. It also has abundant wildlife, including huntable populations of moose, white-tailed deer, blackbean, partridge and excellent fishing.

Access to the area is controlled through 17 checkpoints and access roads, where visitors are were required to register, pay fees for different types of use, and obtain permits for campsites. Visitor days have grown from 121,000 in 1974 to 189,000 in 1984. In 1984, the fees ranged from $2 per day to $17 for an all season permit and were used by the association to construct and improve campsites, run a rubbish collection system and run public education programmes on use of the woods. Though the initial efforts were resisted by those who were accustomed to free, unrestricted access, the less crowded, clean, well-organised system of recreation management is promoting cooperation between landowners and recreationalists. (The Council on Environmental Quality, 1984: 381, 384)

International Paper, North Maine Woods, Inc and other timber companies and land holders in the south and east of the US face less competition from public lands than do timber companies with large land holdings in the west. Companies in the western US must compete with zero prices set for most recreational activities on government owned lands. As a result, timber companies in this region spend little or nothing to enforce their property rights in amenities and tend to ignore these values in management decisions. This is a similar to the situation in Australia, where most native forests are state owned. The creation of an efficient market for recreational and wildlife values in state forests would need to be preceded by implementation of a pricing system on state lands or by the selective sale of forest lands to private interest whether they be industry or conservation groups. Despite the zero-priced competition from state owned forests, timber companies which own forest lands in Australia have provided some recreational values.

Associated Forest Holdings, a division of APPM Forest Products, is Australia's largest private forest-owner. Holdings are primarily in Tasmania and supply 70% of the APPM's regional requirements for timber products. AFH has been following a programme of eucalyptus plantation development since the 1970s and, with a planned 60,000 ha of plantations for pulp production, is a fore-runner in this field. However, the production of timber products has not superseded non-timber values. Forests are managed according to a plan which alternates cutting between selected coupes and leaves wild-life corridors and sanctuaries. A small area of virgin rainforest has been declared a reserve in perpetuity and access to the forests for local fisherman and bushwalkers is allowed on a permit basis (Higgins, 1990).

CONSERVATION GROUP PERSPECTIVE

Efforts by conservation groups provide a different twist on markets in that they are used as a means to protect the environment rather than make pecuniary profits. Indeed such examples run counter to the assumption of some critics that voluntary markets will not work because the natural environment forever remains a 'public good'.

Using primarily volunteer initiative and private funds, private organisations have grown rapidly in the US during the past three decades. In 1950, only 36 conservation organisations existed in the US, but by 1975 there were 173 and by 1982 there were 404 groups representing over 250,000 members. Local conservation organisations in 1982 controlled over 675,000 acres of valuable resource lands (more than 60% of it was in the New England and Mid-Atlantic states, where private land ownership dominates).

Land conservation trusts are generally established with tax exempt status for the purpose of preserving land for its amenity values and for keeping land in agricultural uses. Funds are raised by soliciting members who pay a small fee per year and by soliciting grants from foundations and corporations. With these funds, land trusts can purchase land in fee simple title or purchase conservation easements. These organisations have an incentive to charge fees because the revenues can be used to further their conservation efforts.

In sharp contrast, government resource agencies fail to capture the returns that would be available if they charged realistic user fees for recreation on public lands, and thus fail to protect or enhance recreational values. Speaking for the Trustees of Reservations in Massachusetts, Gordon Abbott Jr. states that:

We are also fortunate that user demand enables us to raise 35% of our operating income from admission fees and that these can be adjusted within reason to catch up with inflation. We're great believers in the fairness of users paying their way.(Abbott, 1982)

At the national level in the US, The Nature Conservancy leads the way in private land preservation, being responsible 'for the protection of 2,916,819 acres in 50 states, in Canada, Latin America, and the Caribbean.' (Dodge, 1987: 2) The Conservancy is also a pace-setter in innovating ways to raise money to cover operating expenses on each preserve it runs. On the 13,000 acre Pine Butte Preserve in northwestern Montana, for example, Conservancy co-managers Dave and Cindi McAllister offer nature tours through the last lowland grizzly bear stronghold in the lower 48 states. They oversee cattle grazing on select areas of the preserve where grazing fees netted $10,000 in revenue in 1986. In addition, the McAllisters started a guest ranch business offering guided nature tours and access to hiking trails, fishing, and horseback riding. The revenues from the ranch help offset the cost of operating the preserve.

Although such a market orientation has not as yet been taken up by many forest conservation lobby groups in Australia, an Australian precedent for such an approach can be seen in the purchase of buildings with historic value by the National Trust. This private organisation funds its own purchases of buildings with conservation value. Once purchased they are managed by the trust on an on going basis primarily for their conservation not their rental value. Nevertheless the trust can benefit from such rents when occupancy does not conflict with preservation. Purchasers' of forest lands for conservation value could easily follow a similar approach.

Private initiatives in conservation can be found at the more local level too. In South Australia a large landowner, Mr Tom Brinkworth, has established several areas of wetland in association with other local property owners. These are financed by charges ($20 per gun) for private duck shoots. Revenues are paid into a 'Wetland Trust' to enhance existing wetlands and create others (David Hawker MP, 1990, pers. comm.). Rather more ambitious is a $15 million venture being planned by Mr Barry Cohen, the former Minister for the Environment. Mr Cohen intends to open an environment park on 170 acres of native bushland at the Hawkesbury River and on the central highway. The plan is to have a sanctuary for animals, giant aviaries to allow free flying birds, an artificial wetland and enclosures for some animals like koalas and possums and for some reptiles. The project is to be financed by a charge of $12–14 per head and by kiosk and craft section. Some conservation groups have also taken initiative in purchasing areas of conservation value. The Tasmanian Conservation Trust recently purchased a private block of native forest in the Tasmanian midlands. (Kevin White, North Broken Hill Ltd., 1990, pers. comm.)

A more colourful example was reported in the Australian in March 1990. Mr Lloyd Bird purchased a former brick-works 30 years ago and has now converted it into a into a miniature rainforest. About 40 endangered species are included in his inventory and some have now grown to 30 metres in height. European pharmaceutical companies have contacted Mr Bird with requests for plant species that could help in medical research (MacArthur, 1990).

Concluding Comments

A carefully structured move towards a more market based forest industry has the potential to achieve an improved balance of environmental and economic goals. In the 'ecologically robust' eucalyptus forests, market based management, for timber production alone, would create a more efficient timber industry, with less wasteful investment in the native forest, a better timber harvesting decision, and possibly a reduced area of logging. Extending the role of markets into the less traditional area of providing amenity services in the forest has the potential to improve the outcome still further.

State forest services in Australia currently provide non-wood forest values at a zero price. Queuing and rationing results. Market provision of such values is under-cut and out-competed, so there is little incentive for either public or private resource managers to provide non-timber uses such as wild-life and recreation. Because rights to forest resources are publicly held and not transferable, conservation groups do not bid for forest lands and have no incentive to consider timber values in their political vote-seeking.

Under a market alternative, forest lands would become saleable with both the timber industry and conservation groups entering the bidding process. Examples from the US, such as the International Paper Company and The Nature Conservancy illustrate how this can lead to provision of more amenity values. The Australian examples cited illustrate a local capacity for similar developments, but government subsidisation of recreation and environmental amenities under-cuts this potential for private provision. Selective sale of Australian native eucalyptus forests would encourage more private provision of conservation services and values, without the extreme of a total moratorium on timber production. The presence of 'free-rider' problems means that the outcome still may not be the theoretically optimal trade-off between wood, wildlife and wilderness. In the absence of detailed information about peoples preferences, political allocation of the forest resource also has real deficiencies involved in it; and it is our view that the problems of political failures are far more significant than market failure. Market based allocation at forest resources may not be ideal or perfect but there appears to be no superior alternative!

6 Mining and the environment

Executive summary

Minerals and mineral processing are major fields of Australian economic activity. Australia clearly has a natural advantage in this area and the energy of its explorers, miners and processors has supplemented this natural advantage so that the industry today is a world leader. Mineral and processed mining products account for about half of Australia's exports, and mining comprises 7% of value added in the economy. Australian mining industry output has shown an eight-fold real growth in the past quarter of a century, with output surging strongly in the late 1960s/early 1970s and again after the early 1980s (see Chart 6.1).

Australia's prominence as a producer ranges across a wide variety of mineral and energy products including coal, iron ore, bauxite, gold, natural gas and copper lead and zinc. We are also the world's most important supplier of mineral sands and in recent years have become a leading producer of diamonds. Although government restraints have prevented Australia becoming a major supplier of uranium, located within the country are some 30% of the world's low cost reserves.

Exploration expenditure in Australia is significant on a world scale (see Chart 6.2).

Environmental and sustainable development issues stemming from mining can be divided into two parts. The first involves the direct and intentional effect of the activity—the management of a potentially depletable resource. The second is at the heart of the matters examined in this book and concerns the potential for mining activity to generate pollution and land degradation.

Management of a depletable resource

Man neither creates nor destroys matter.

No metal is ever depleted in the sense of becoming irretrievably changed and no longer available. Metals are *modified*. They

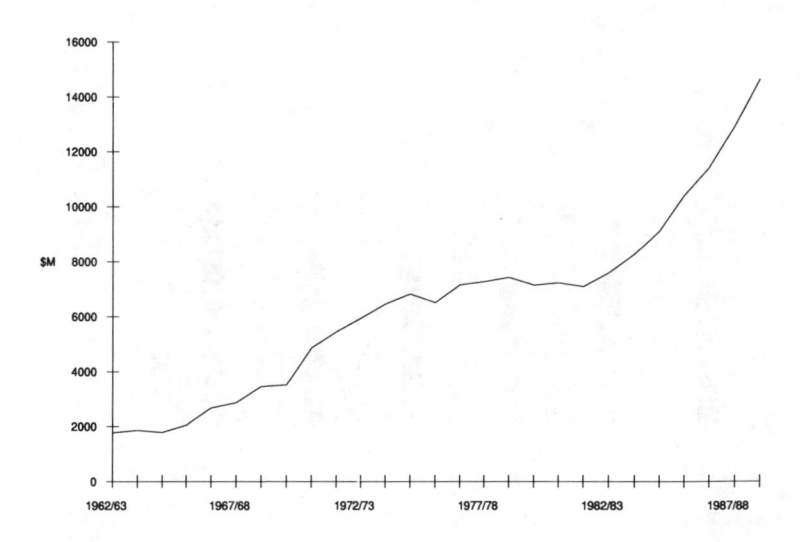

Chart 6.1
Gross Domestic Product – mining (1979–80 prices)

are extracted from their location, used elsewhere and—at least in principle—are available for continual re-use, through recycling. Where recovery is more economic than the development of virgin resources, it will be the preferred approach. More commonly, recycled materials will be used alongside virgin extraction as both have highly variable extraction costs. Low cost sources of both types will be used first.

Energy materials, the 'banked' reserves of the product of solar radiation, *are* used up where their burning transforms them into energy and residual matter. But this too is simply a transformation of their components back into a form of their original state.

Choices confronting a mine owner

Owners of a depletable resource confront choices similar to those facing owners of a renewable resource. In both cases, the objective is to maximise its value. For the owner of a renewable resource, the decisions encompass trading off expenditures which assure a future income stream against the expected value of that stream. The owner of an individual ore deposit does not have the

Chart 6.2

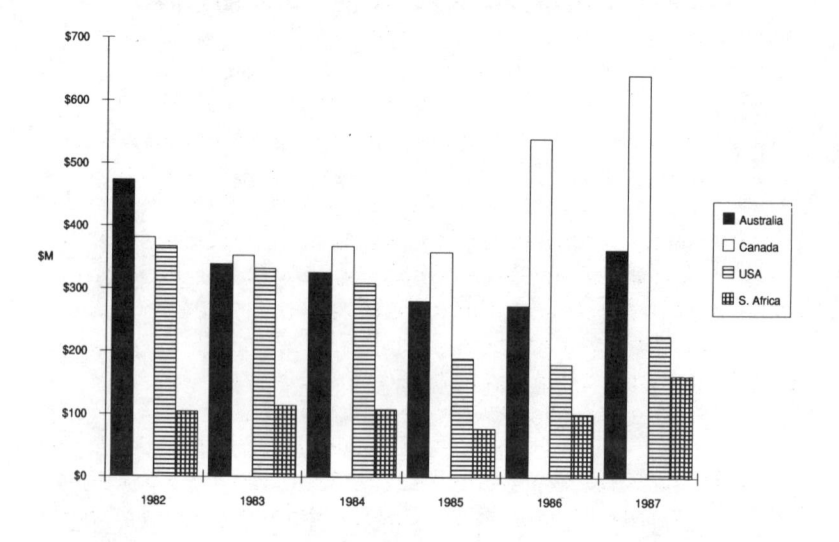

Source: CRA Ltd.

option of maintaining it in perpetual use, still less the possibility of its future use being made more productive. Nonetheless, the owner of the ore deposit will take similar steps in weighing up the merits of the pace of development and in evaluating future strategies on capital expenditure—including searching for new resources—against simply consuming the income derived.

Robert Solow, in his 1973 Ely Lecture to the American Economic Association (Solow, 1973), noted the thinness of economic analysis of resource depletion at the time. He chose that subject for his own address because of the outbreak of concern spawned by the Club of Rome's 'Limits to Growth'. In doing so, he drew heavily upon Hotelling's 1931 article, *The Economics of Exhaustible Resources* (Hotelling, 1931). According to Hotelling, a resource must earn the same net income return as others in its risk class. This may be achieved by either extracting it today, or leaving it in the ground and allowing its capital value to increase prior to extraction at some future date. For the owners to leave it in the ground means taking a view that the net returns will increase at an exponential rate at least equal to the rate of interest.

It follows that if the rate of interest increases, or if there is an expectation that future price levels will fall, extraction rates will increase. For the resource owner, one nightmare is to be left holding a resource after it has been technologically superseded. Thus, at the time of the discovery of gas in the Netherlands during the 1960s, all parties were anxious to expedite its extraction in the expectation that nuclear power would mean its value would fall. Similar concerns are a part of the Saudi oil production strategy.

Solow drew attention to an empirical ambiguity in this relationship, namely that when prices were generally expected to fall, owners would increase their production rates with the extra supply thereby bringing a marked reduction of existing prices. Conversely, when future prices were generally expected to rise more rapidly than interest rates, supply would be constricted and present day prices would rise. Noting that chronic instability of this nature did not occur, he introduced the potentiality for asset revaluations. Existing reserve stock values will adjust up or down in response to changing price expectations. The 'flow destabilisation' is thereby offset.

More important than the sleight-of-hand explanation involved in using asset revaluation is the cost side of the equation. Where increased output requires capital expenditure, higher levels of interest also add to the costs of extraction. The relationship between the two forces will depend upon the present value of expected resource rents, and the cost of the additional capital requirements.

Stollery (1990) conducted a simulation exercise on copper and US eastern bituminous coal and concluded that real pre-tax discount rates of 9–10% were required before the depletion rate is accelerated. He noted that these levels of return are similar to those required of mining investments and hence the discount rate is approximately neutral in its effects on the rate of depletion.

Influential though Hotelling's analysis may have been to economists, it rarely finds a place in mineral development decisions. It would take a firm or nation with unusual market power (as De Beers has in diamonds or as Saudi Arabia may have in oil) and with considerable confidence in its foresight, for it to be able to forgo the development of a profitable field in the expectation of a price increase. There is rarely a universal view on whether prices are set to rise or fall—at least a view that is sustained for any length of time. Moreover, there is no uniform cost profile of mining activities. It may be that assets are revalued in response to expectations, but extraction costs will vary markedly between mines and deposits

Unless a firm has market power in supply of a mineral, it would be pursuing an extremely high risk strategy if it were to sit on a resource it could profitably mine in the expectation of prices rising. Sitting on a genuine mining prospect means deferring income from something found in only one in a thousand search attempts. It would be even more risky to adopt such a strategy in the expectation of prices rising faster than the prevailing rate of interest. In the main, real prices of resources have fallen, notwithstanding lower grades or less accessible reserves progressively becoming available. This fall is a function both of market demands (greater technical efficiency of use for basic materials and higher income elasticity of demand for services rather than goods) and improvements in extraction technologies.

It is, of course, a different matter to defer development of a marginal find in the expectation that improved extraction techniques will, at some future date, make it more viable. Where technology changes are bringing greater falls in extraction costs than the price falls caused by action on the demand side, deferral may make good business sense.

Sustainability of mining

In principle, a nation's mineral resources are depletable in much the same way as those of an individual mine. As discoverable resources are ultimately finite, there is a potential case for society placing a limit on expenditures allocated to searching. Indeed, without a limit it is arguable that excessive expenditures will be incurred during a scramble to prove and take up resources. Such a situation would be akin to having no mining lease available in a highly prospective area and has been known to occur especially in alluvial gold recovery. When reserves are large or unknown, limitations to search activity would be inefficient.

In its submission to the Industry Commission, the Australian Conservation Foundation (ACF, 1990) argued that ecological sustainability was a pre-condition for the maximisation of consumer welfare. The ACF quoted extensively from Dr H. C. Coombes, who draws attention to what he regards as the finite nature of mineral resources and 'the appalling history of the mining industry in its damage to the environment'. Although they also drew on Hicks, (who saw the need to substitute renewable for nonrenewable resources when the latter were used up), the ACF's focus was on nondevelopment. They defined ecological sustainability as '...the passing of the natural environment from one generation to the next in a condition relatively unaffected by human activity such that the ability of future generations to provide for their own needs is not compromised by the present generation.' It should be noted that this definition follows the narrow view of sustainable development favoured by Pearce et al. (1989: 48): and not the broader view, which they also found acceptable, which allows substitution between man-made and 'natural' capital.

In strict terms, it is very difficult to see this narrow view as being compatible with any mining whatsoever.

A more measured view was presented in the Brundtland Report (1989), which it focussed upon use of non-renewable resources in reducing the stock available for future generations and added,

> But this does not mean that such resources should not be used. In general the rate of depletion should take into account the criticality of that resource, the availability of technologies for minimising depletion, and the likelihood of substitutes being available. ... With mineral and fossil fuels, the rate of depletion and the emphasis on recycling and economy of use should be calibrated to ensure that the resource does not run out before acceptable substitutes are available. Sustainable development requires that the rate of depletion of non-renewable resources should foreclose as few future options as possible. (Brundtland Report, 1989: 45)

The processes envisaged by Brundtland are in fact precisely those performed by markets. A rising need for a resource increases its price. On the one hand this stimulates the search for more of the resource and for substitutes for it; at the same time, it provides incentives for dispensing with or economising on its use.

The quantum leaps in oil prices in 1973–74 and six years later brought an intensification of exploration, a substitution into other fuels (one which was severely hampered by restraints on nuclear power station building in most countries) and greater economy in fuel, especially oil, usage.

The upshot was a shift in the relationship between energy and GDP in OECD countries from over 1:1 to 0.7:1 and of oil from 1.6:1 to 0.4:1. The amount of energy required to produce a dollar of US GNP has fallen 28% since 1973 and further savings in electricity usage are underway (Ficket et al., 1990)

Much of the history of mining has been a saga of shortage driving up prices followed by glut as both supply and demand have responded. Furthermore, we should not be overly concerned that future emphasis of this process will be on the demand side as a result of previous depletions of resources.

It is our own view that resources are best regarded as being infinitely available. Substitution and discovery of new reserves has meant that no resource is approaching depletion. Even if it were, the appropriate framework would be to arrange for its depletion to be allocated over a time path rather than assume it to be more valuable in the future.

On a plausible assessment of the distribution of minerals around the globe and of the costs of recovering these at current growths of usage, no metal (or coal) will need to see its production reduced over the next 300 years (de Vries, 1989). And those metals which might be approaching such a position—copper, lead and zinc—have ready substitutes. The most important metals— iron and aluminium—could never be mined out.

We have inadequate information on the limits of unmined mineral availability, and of the value of that availability in relation to the costs necessary for its discovery. It follows that there is no case for limiting search activity. Decentralised market forces, through the pricing mechanism and interest rates, provide sufficient signals to explorers when to pare their expenditures.

If prices rise faster than the interest rate, there will be an incentive for known resources to be reserved from present use and increased activity in searching for new sources—the owners are better off holding present resource stocks than developing them. The fact that this so seldom occurs is purely a function of the expectations (which have proven to be well founded) that the pay off would not warrant such action.

The collective wisdom of the marketplace, in its price signalling and its capital demand signalling (interest rates), has proven to offer adequate accuracy as to the ranking of future as opposed to present needs. Where governments have sought to intercede—for example, by subsidising new energy types in the expectation of shortages not anticipated by market participants—the activities have generally proven to be ill-advised.

Similarly, it is difficult to justify government intervention, in expectation of future shortages, to tax or otherwise discourage the search for new mineral deposits. (This is not to argue against charges which compensate for services rendered.) It is even less likely that highly distortive measures like export taxes and extraction of additional profit taxes by monopoly rail freights would contribute to efficiency and sustainability.

Property rights and efficient resource management

The basic principles on how policies should be formulated to bring about efficiency require secure, privately vested property rights, together with government allowing markets to operate freely, and occasionally taking steps to facilitate their operation. Where things cannot be individually owned, charges should be introduced so that their use can be allocated among competing needs. In these respects, secure vesting of property rights need not mean that everything must be owned. It is not necessary, and might indeed lead to inefficiencies, if things without scarcity value are required to be owned. Nor should charges be levied on things with no scarcity value.

Within the framework of the rule of law, individual ownership ensures that adequate care is taken of property and that, where its use imposes costs on others, redress is available. Ownership creates efficiency. Inefficiency occurs where things are unowned and users lack incentives to allow resources to be used for the purposes which offer greatest value. With an ability to trade, resources are generally drawn to their most socially useful purpose.

Where assets are not presently vested in firms or individuals, there is a powerful case that they should be given—or sold—to the party valuing them most. This, of course, may well have equity implications but the key to efficiency is individual vesting and greater efficiency gives capacity to better afford goals like income redistribution.

Higher income levels are commonly designated as economic goals which, together with environmental goals, form part of the well-being which each of us seeks to maximise. However income is too narrow a definition of economic goals and the efficiency stemming from market-based instruments, like property rights and charging mechanisms, extends beyond the generation of income. Private property rights and charges for using common property will also foster the trade-offs between alternative uses of resources, which individuals (and, as a result, the community as a whole) wish to see. Foregoing any use is among these alternatives. Market-based instruments also offer incentives to bring about wider goals like the avoidance of both waste and environmental degradation. They do so through two mechanisms: providing a basis by which conflicting needs can be traded off; and improving overall wealth levels which allows more of all needs to be satisfied.

Just as there are trade-offs between work and leisure in achieving this well-being, so there are trade-offs between environmental goods and income levels. Moreover, just as higher income levels offer us a greater ability to afford increased leisure, those higher income levels also offer us greater capacity to afford environmental goals.

The issue is how we might best optimise income, leisure, environmental and other goals. The keys to this process are founded on a comprehensive vesting of property rights, whereby owners seek to maximise the value they obtain from their assets, while providing adequate compensation for any costs they impose upon others. This process has a relevance way into the future. The trade-offs between uses allow resources to be held back from present day usage in the way Hotelling posited. If it seldom pays to reserve goods from use in the present or near future, this indicates that owners have little confidence that they will have a higher value in the future.

As has been previously discussed, that lack of confidence is firmly based—the experiences of the past tell us that technological advance has resolved what might have been shortages. Whether or not markets will prove accurate in the future, we have little alternative but to allow them to remain the arbiters. It is doubtful that governments can claim a greater prescience than individual owners about the future worth of resources, if only because they have a diminished interest in obtaining maximum value from those resources. Governments are, moreover, driven by political contributions which may well override the efficiency generated by markets arbitrating between willing buyers and sellers.

Ownership of minerals as a means of promoting sustainability

Although it is convenient to discuss property rights as though they conform to some universal definition, this can be misleading. There are many varieties of property rights, none of which confer absolute jurisdiction. In fact ownership of property simply signifies access to a number of service streams which might derive from an asset, together with obligations for certain adverse impacts the ownership of that property might generate.

Rights accrue with or without a formal title. Some of these rights are spelled out in statutory law, whilst others are left to common law. Thus, where a mining company undertakes authorised work on land owned by someone else, it obtains certain rights to exclude others (including the owner) from making use of that work. Whether as a result of law or tradition (and the two notions converge), legitimate claims or benefits are generated.

There are things for which a case for vesting can rarely be made. These include solar radiation, which is infinitely available—its use by one party seldom detracts from the value of the residual amount available to any other party. Widely vesting rights to solar radiation would simply create transaction costs, which would reduce usage of an infinitely available resource.

Solar radiation is an extreme case of an abundantly available good with no exchange value. Fisheries offer a contrasting case. Most fisheries are unowned and have a positive value; but with limits on the ability to exclude anyone from participating in the catch. This lack of vesting has led both to excessive effort being employed in taking the catch, and to depletion of the resource because each fisherman attempts to maximise his catch without regard to the on-going value of the fishery. Where rights to fish can be defined and protected, depletion will occur only when this is the rational outcome for society as a whole.

A mineral lease performs similar functions. It gives the miner the right to extract value at the time of his choosing (unless other arrangements are contracted for). In many respects this is similar to fishing rights, although the latter give the right to 'mine' a specific quantity of fish within a designated time. With this caveat, delineating a mineral lease is akin to granting rights to a fishery. The owner is given exclusive rights and incurs expenditures to prove the resource so that the expected value, less the expected costs, maximises the income available. Unlike a fishery, the development process of generating income from a mine is likely to follow a plan involving depletion of the resource.

The initial value of an exploration lease is likely to be very low; but mineral reserves, once discovered and delineated, do have a high value. The primary issue is to devise means by which the appropriate amount of effort is allocated towards searching out potentially valuable unknown resources. Part of this involves establishing rules by which such valuable resources may be developed (or be maintained for other uses should search reveal value in their not being developed). The greater the likelihood of successful search activity being followed by denial of subsequent development, the less the incentive to conduct searches. Expenditure on search activity will be encouraged when there is certainty and clear rules on when a successful search will be permitted to proceed towards development.

With the foregoing qualifications, in the development phase

itself, economic forces in free-markets automatically offer an efficient solution. By this, we mean an outcome which alternative arrangements are unlikely to improve upon. The property owner or entrepreneur weighs up the value likely to be received against the costs likely to be incurred. To do so, the party undertaking the expenditure must be reasonably assured that it will obtain any profits likely to emerge. With secure property rights, as long as the resource is definable, excludable and tradeable and as long as there is adequate competition, the appropriate amount of effort will be allocated. Where rights do not conform to these criteria, either too little or too much energy will be expended.

Allocation of exploration and mining rights

All rights to minerals in Australia are vested in the crown. Mining leases are allocated on a 'first come first served' basis. At first blush, the best market-based approach to the allocation of mining rights would appear to be 'long tenure', unconditional and tradeable mineral rights vested by competitive cost bidding (alone or subject to pre-announced royalties).

The deficiency of this approach is that competitive auctions usually reveals information to the benefit of the passive follower and at the expense of the active discoverer. In a sense we have a system of auctioning at present—there is only one applicant for the great majority of exploration leases.

In the main, the community obtains most benefit when its institutional arrangements allow firms to allocate resources they see as being optimal to the discovery of unknown value, rather than arranging for the assignment of value which is relatively certain. Furthermore, the situation in mining is dominated by the goal for the discovery of value. This requires active steps.

Accordingly, it is to the benefit of society if institutional settings are in place which do not impede the search for value. Long tenure and unconditional rights would favour passive waiting for others to discover value. While this might be appropriate where future uses of resources are reasonably certain, the fact is that mineral deposits in Australia remaining to be discovered are totally unknown; and, to all purposes, it is as if they are infinite. As previously discussed, price and substitution responses are such that no particular resource will ever run out. Government intervention to conserve exploration effort is therefore not warranted (and even if it were, government neither has the information nor confronts the incentives appropriate for it to arrange the correct conservation strategies).

Hence, the present system of 'first come, first served' is an efficient method of ensuring that the right incentives to search out value are put in place. By imposing disciplines on explorers in terms of the requirement to progressively surrender parts of their leases, or through systems of work program bidding requirements, the search for value is stimulated. The community as a whole obtains a share of the successful searches for value, both indirectly through increased activity and directly through the payment of royalties. The latter, if efficiency is not to be unduly impaired, should be at a pre-arranged level.

In respect to efficiency, the nexus between exploration and the right to mine is crucial. The value of the site is created by the skills or good fortune of the explorer. Measures which are likely to introduce risks that discovered value may not be claimed will rebound on the incentive to search. This adverse effect on search activity will be greater, given risk aversion, where it is less certain or arbitrary than where it is known in advance. Thus, if the government is to demand additional revenues amounting to 20% of the profit, and if each exploration attempt has similar costs, there will be a reduction in search expenditure of 20%. If, on the other hand, the explorer cannot easily determine what additional share of revenues must be yielded, he is much less likely to risk any of his capital.

The Industry Commission (1990) favoured cash bidding for successfully explored land and suggested that rights to mine a promising exploration find should be tradeable. Tradeability of rights is, however, the present position. A wildcatter, or even a major company, will often seek a partner to develop a promising find. In doing so, the value of the successful exploration will be sold (perhaps even auctioned), so that the discoverer will receive his reward. The Commission's suggestions that a discovery should be re-auctioned were focussed on the taxation and allocation aspects. Yet, no explorer will reveal all the data necessary for others to make fully informed judgements, and a requirement like this will generate strategic rather than wealth maximising behaviour. Attempts to police these various strategies would involve the government in considerable expenditures and a raft of black letter law, similar to that governing takeovers, with an inevitable growth of bureaucratic monitoring.

Cash bidding has a place in some circumstances. But auctions are likely to be the most efficient means of allocating tenures only where prospectivity is known to be high, as in some petroleum rich provinces and in certain coal areas. The original discoverers of coal in Sydney Basin or oil in Bass Strait could not reasonably expect to have a lien on future discoveries beyond areas to which they were granted exploration rights. Their discoveries revealed value which they could not quarantine to themselves. In such cases, allocation by cash bidding is likely to avoid wasteful rent-seeking activity and, perhaps, premature development expenditure. But these situations are rare.

Systems like Australia's crown ownership of undiscovered minerals carry considerable advantages for efficiency. Government ownership is likely to bring adverse effects, where the intent of this is to limit or ration access; but such ownership in the Australian context may best be thought of as representing custody, pending assignment of relevant rights to those who can demonstrate value from having such an assignment. In this respect, the model to which it bears closest similarity is the ownership of western lands by the US Federal Government during the nineteenth century. Title to those lands was vested in the Federal Government with the intent that they be placed in private hands at an early opportunity (Anderson and Hill, 1990). The US government, rightly, sought compensation from the private landowners in part to arrange efficient allocation and in part as compensation for its custodial activities.

The difference between the model applied to the US western lands and that which we see as applicable to mining largely involves the custodian's requirement to a separate usage fee, in the form of a royalty, rather than to up-front sales revenues (which in the US were sometimes waived). A further difference is that economic rent, or income in excess of that needed to warrant production, was present in the case of land settlement; whereas this is mainly absent—or must be created—in the case of mineral exploration. We shall return to the point presently.

Compared with land being held in custody by the crown, permanently vesting ownership to minerals (either in the surface owner or as a separate title) offers inferior incentives to the search for value. Ownership of minerals by the surface owner, aside from generating arbitrary and inequitable windfall gains, appears to yield much reduced incentives for explorers to search out value. Although comparisons can never provide watertight evidence, the contrasting exploration expenditure outcomes in Ireland (where crown ownership prevails) and England (where the land owner has the mineral rights) offer persuasive evidence of this. Similarly, there is much less exploration activity in the eastern part of the US than in the western states where government ownership is prevalent.

Vesting ownership of minerals in the surface owner would simply introduce transaction costs whilst conferring little in terms of incentives to search out, improve, or guard the resource; the surface owner has no idea that the resource exists (and for the overwhelming bulk of land, no valued mineral resource does exist). Vesting ownership in this way would be similar to assigning ownership to the electromagnetic spectrum prior to any use being found for it. In that case, prior assignment would have offered something of value to the owner where the worth was totally unknown, and would have resulted only in impeding the discovery of that worth.

The reduced scale of operations observed where mineral rights are vested prior to value being discovered, or in the absence of any intent to conduct search activity, also often arises due to additional transaction costs likely where miners/explorers must deal with a variety of owners. Some owners may choose to 'hold out' and embark upon other strategic behaviour.

Assigning ownership where no value has been demonstrated would also introduce a new set of uncertainties. The prospects of governments reneging on assignment are further likely to undermine the sums which a prospective owner of mineral rights might be prepared to outlay. These sovereign risk uncertainties would not be seen to be diminished if the rights were to be vested. The owner would never be certain whether the government might take steps to expropriate a highly valued income stream.

Moreover, it is by no means clear that the sale of the rights would yield significant revenues. Land, outside of certain highly prospective areas, is mainly valueless as a source of minerals. Intellectual property rights, that is, knowledge about deposits, are typically the scarce resource, not the physical deposits themselves. This distinction between physical and intellectual properties has a major bearing on the taxation of mining because it means that there are rarely any real economic rents accruing to the activity; there are simply high profits to those firms which, purely through the skills they have developed, economise in their search:discovery ratios. That is, they use their expertise to conserve outlays to those areas where efficient and creative research has revealed prospectivity.

Mineral taxation and sustainability

Much of the literature on mineral development examines policy approaches from the perspective of taxation. General principles of taxation, particularly neutrality, are grafted on to two apparently observed phenomena which mark mining as different from, say manufacturing. These are that rents, or super-normal profits, are apparently observable from many operations; and that the mineral resources, in Australia and elsewhere, from which these are derived are the property of the people as a whole.

Models of taxation to ensure an equitable and non-distortive allocation of rents to the people as a whole have been developed. The least distortive of these are based on the work of Brown (1948), which, though not directed at mineral taxation as such, has been widely applied to it. Essentially, a pure Brown tax would see the government sharing in both the costs and profits of a venture. Variations on this have been put forward by the Industries Assistance Commission (1976), Swan (1976) and, in a distorting form which seeks to ensure that the government will share only the successes, by Garnaut and Clunies-Ross (1975).

In all these cases, the preferred approach is posited on the existence of super profits. Some have noted that auctioning would offer a superior method of obtaining the rent, but auctioning is seen to be deficient in view of collusion. There is some unease with the proposal to extract rent tax expressed by, for example Smith(1979) who observes that exploration is likely to be risky and should be rewarded accordingly. Church (1985) notes that coal deposits are widespread throughout the world and its mining technologies are well known, so that the existence of high profits would be expected to be competed away. Notwithstanding this, he cites econometric evidence which appears to demonstrate that rents in fact exist; but his explanation of their existence (a limited field of potential bidders) does not seem entirely plausible.

Whilst it is unquestionably true that high profits are available from mines in specific locations, this is a far cry from the common description of rent. Those high profits are created through knowledge related and entrepreneurial processes, rather than being pre-determined. Striving for this form of rent involves a continuing search for highly profitable activities, and is the essence of a dynamic economy. As Schumpeter said in describing the process of income creation:

> .. competition from the new commodity, the new technology, the new source of supply . . . which commands a decisive cost or quality advantage . . . is so much more effective than (price competition) as a bombardment is in comparison with forcing a door, and so much more important that it becomes a matter of comparative indifference whether competition in the ordinary sense functions more or less promptly.

(Quoted by K.P. *Barwood Implementation of a Competition Policy* NZ Association of Economists Conference, February 1986).

The entrepreneur searches for hidden values and may 'create' needs in the hope of earning high rewards. Where these rewards are attenuated by taxation, the search for them will be so much diminished. In a sense the finder of a high yielding deposit has a monopoly on its exploitation and sells its outputs at the best price he can.

The notion of undeserved economic rent accruing to mining companies has dominated many economists' thinking about this industry. While individual mines and companies earn high profits, economic rent which can be taxed without affecting production decisions is rarely observable in mining activities. Most commonly, rents are simply the occasional high profits which result from corporate skills and research and development activity. They are, in short, earned rather than given; where their potential existence is observed, this is typically followed by their 'dissipation' in further research and development. In mining, as with other industries, such 'dissipation' leads to discoveries of value. And, although in principle it might be possible for the nation as a whole to maximise net revenue by economising on the R & D effort, in practice this seems unlikely.

The lack of aggregate rents accruing to the recovery of minerals can be corroborated by examining the profitability of mining companies. Taken as a whole, mining companies' profits are similar to those of other companies. In its submission to the Industry Commission, AMIC pulled together figures which showed that, over the period 1972–73 to 1986–87, mining returns on capital were in fact 11.7% compared with 14.7% for all industries.

If aggregate profitability in the mining industry *were* to be somewhat higher than in other economic activities, this would be a reflection of risk premiums rather than of the existence of (Ricardian) rents. If rents were available or became available, entrepreneurs would switch out of other industries to find them.

We should therefore recognise royalties as a form of additional taxation, legitimised both by centuries of operation and by notions of providing a charge for government custody of resources. This latter element would be quite a small proportion of the monies actually diverted to governments. There is no case for royalty collecting in excess of sums needed to finance wardens' courts, government geological mapping, etc. Beyond these levels, except for royalties imposed to allocate reserves between rival firms, they will normally be forms of taxation which discriminate between economic activities and, therefore, reduce aggregate income levels.

The exceptions to this—areas of high prospectivity and strong commercial rivalry for exploration leases—may be thought of as having changed the nature of their tenure. Tenure of such highly prospective leases is best regarded as having shifted from government *custody* to government *ownership*. Information on the value of the areas has become so widely known that a system of 'first come, first served' would prove to be wasteful of services and reveal very little additional knowledge. When economic rents are known to be present, Government ownership, rather than custody, become the efficient form of tenure.

Environmental protection and mineral extraction

In recent years, mining operations have attracted considerable opposition, especially from environmentalists concerned at the destruction of extant 'natural' areas and pollution externalities.

Environmentalists' concerns about mining are on a stronger theoretical footing when they address externality issues than where they seek to interpose restraints on development to pursue sustainability. Although many externalities are trivial and not worth the transaction cost or effort of arranging for compensatory payment, others may lead to distortion by inadequately valuing the spillover costs. There are two issues in this concern: the excessive resources attracted to activities which have their costs partly born by non-contracting parties; and the unknown risks of development.

With regard to the latter, all activities carry risk; but through well constructed liability laws, proper incentives are put in place, so that parties balance their gain against the costs. These costs include a premium to account for risk. The party most directly involved is better placed to take such judgements than a remote but 'central' authority. The party directly involved is likely to be more familiar with the terrain and the various techniques available to reduce exposure to mishap, as well as having a direct stake in the downside and upside rewards. Such a party is able to weigh up the value obtainable for itself (and indirectly the community) against the risks involved.

Incremental judgements by independent market operators

may also impose less risk than centrally determined judgements with a wide application. More importantly, where government agencies take decisions, they are likely to be excessively risk averse, since they have no direct stake in the rewards available. Hence their actions are likely to bring opportunity losses to the community from options foregone.

Internalising externalities is always an attractive approach both to address known polluting activity and to ensure that risk is taken into account. The externalities imposed by mining activity itself include destruction or impairment of wilderness or other scenic areas through:

- scarring the landscape;
- harming rivers and other pre-existing systems;
- reducing the value of land for purposes other than mining.

Land as a commonly owned resource and as a bequest value

Where property is fully assigned to individual ownership, and where externalities are not significant, it will find its uses allocated to the purposes which the community values most highly. Because of the difficulties of totally eliminating externalities, some land is reserved from uses (farming, mining) which would fundamentally alter its present nature; and even from uses (tourism, bush-walking) which would alter it less significantly.

The justifications for such reservations find expression in 'existence' and 'bequest' values, which, it is argued, could be even less comprehensively secured by market signals than land held back solely for its leisure uses.

For these reasons there is no constituency for fully privatising all land—to most people the risks are too great that there will be an inadequate capture of all the 'non-use' values and that more readily tapped private purposes would usurp prized recreational, 'bequest' and 'existence' values. In part this fear is likely to be compounded by a feeling that fully privatised land would leave those with powerful lobbying muscle worse off, in that the land use would shift to purposes which they do not favour.

Accordingly there is wide support for land being reserved in national parks. However, there is no yardstick, once allocation by means of commercial market processes has been abandoned, by which we are able to measure the appropriate amount of land that should be so set aside.

All mining is effectively banned in national parks and, indeed, even aerial exploration is forbidden. Hence we are not only denying ourselves opportunities for wealth generation but also preventing ourselves from becoming familiar with the possibility of even knowing what these opportunities are. It is as if we cannot trust ourselves even to look at the cookie jar. Such restraints on scientific curiosity sit uneasily with our wish to enhance our well-being.

Beyond national parks, the ability to take up leases has also been progressively restricted in a great many areas. Thus, as a result of delays by cascades of approval procedures, Roxby Downs, located in an area of no value as wilderness and of marginal farming use, has required 54 separate approvals from regulatory authorities. This process has occupied the best part of a decade, and is in contrast to the situation prevailing during the 1960s when the massive Kambalda nickel mine required just two separate approvals.

None of this is to argue that the decision to allow exploration should automatically be followed by the right to mine. The exploration activity itself may reveal values that the community may wish to use differently and thereby forego the benefits of mining. Alternatively, the expected values the exploration reveals may be insufficient to justify disturbing a particularly rare ecological formation. However, government should be wary of attempting to involve itself in detailed cost-benefit analysis as a condition of allowing developments to proceed. It neither has the expertise nor faces the appropriate incentive structure to pursue such a policy.

Considerations about the trade-offs between unpriced and market values are—or should be—more straightforward for a highly concentrated activity like mining than in the case of more land extensive uses. Any good is more highly valued where it is scarce. A children's playground is less valued in an area with a large number of such facilities than in an area where they are few and far between. Because mining in remote locations takes up such a tiny proportion of the area in which the mine is located—typically less than a square kilometre in a one thousand square kilometre exploration field—and because it is a highly valued activity, it has in the past been considered a preferred form of land usage. The various restrictions and impediments imposed on mining indicate that this preferred usage has been modified in a way which is difficult to justify.

Greater income and material well-being certainly allows us the scope for setting aside resources to satisfy other needs, but this has occurred in Australia through government fiat rather than voluntary action of individuals. Notwithstanding that externality and 'free-rider' problems present difficulties in this area, there is a strong likelihood that political log-rolling has brought a far more prominent role for government in excluding commercial activities than would be the wishes of the people if given a true market type of choice rather than ballot box choices, which amalgamate a great many issues which have only a remote bearing on true preferences.

These same considerations apply to the notion of reserving land as a bequest to future generations or maintaining it for its existence value. It is important to maintain a perspective on the overall land degradation brought by mining activities. At any one time, mines account for less than 0.02% of Australia's surface area (AMEC, 1990); and it is likely that less than 0.1% of Australia's area has ever been mined. Of this, some would be economically viable for restoration; other areas, especially those near population centres, would have a value as land fill sites. Still others are more valuable left as disused sites; forming, as they do, a part of the nation's heritage.

Mining and other land use conflicts

The Industry Commission's draft recommendations on public land use conflicts called for rigorous cost-benefit analysis, new land use categories, creation of independent advisory bodies, and a requirement that proposals be assessed from a overall public viewpoint. These recommendations conflict somewhat with those under 'Environmental Impact Assessment', which call for market-based solutions (presumably narrowly confined to those parties having a legitimate standing with regard to the issue).

This latter approach has two advantages. First, it prevents decisions being taken on political grounds. Secondly, it avoids cluttering the process by the intercession of parties with only a remote interest in the outcome. Rather than preview action, the law normally operates by ensuring liability is in place in the event of harm being imposed. This leads to an efficient use of resources, allowing economical decentralised decision making by those with the best information.

Although the former approach was intended to facilitate approval processes, experience demonstrates that creations of advisory bodies, and requiring proposals 'likely to have a major impact' to be examined from the overall public benefit viewpoint, will lead to increased regulation. Taken further, such a proposal may amount to a form of centralised planning whereby the decision to go ahead requires the imprimatur of the government.

Mining is almost always confined to very small areas—the Coronation Hill mine, for example, would amount to the size of Parliament House in an area the size of the ACT. This brings us to the view that the community would be ill-served by a requirement for rigorous pre-assurances that a new activity in mining will not do harm. To assemble evidence for, and conduct deliberations about such reassurances is costly both in time and money.

These activities will most certainly reduce the return available to the community. Of course, where a major development has the potential to impose considerable incidental costs on the community—for example in the case of extensive expansion of farming on previously unused land—more rigorous pre-assessment they may be justified. For mining, experience (and intuitive knowledge of the extent of mines) indicates that applying such an approach means the positive opportunities missed would be likely to outweigh any costs.

Rather than supporting measures which would bring about delays (like those the Industry Commission recommended under 'public land use conflicts'), it is preferable to devise measures which facilitate resolution between the parties most directly affected. This would be further assisted by placing strict time limits on approval processes both for mining and, where mining might not readily be permitted, for exploration.

Pollution issues

Notwithstanding the general acceptance in principle of Coase's proposition that environmental impositions are mutual issues not necessarily requiring compensation or mitigatory action on behalf of the 'victim', the conventional approach has followed the Polluter Pays Principle. Issues concerning pollution are more fully addressed elsewhere in this volume. In mining, because of the localised nature of operations and their highly concentrated nature, pollution is more straightforward to monitor; simpler to contain; and (where the operation is remote) easier to dilute so that it becomes relatively more harmless than is the case with less compact emission sources. Australian mining companies have accepted the Polluter Pays Principle and the issue largely revolves around how, and how strictly, it might be applied.

Important in this respect, the ability to rectify mistakes and resiliency is demonstrably the case in mining. Indeed, nature itself, in areas of greatest environmental sensitivity, has performed this function, for example, in Kakadu gold mines, where cyanide treatment was used without any precaution against environmental damage. Arguments for Kakadu to be maintained in its present primitive splendour sit uneasily with the fact of previous pollution.

Rectification of land

It has been suggested that mining has in the past left a legacy of degradation and destruction on the land, and that measures should be enforced to prevent this happening in the future. Evidence of these unfortunate occurrences is seldom presented— or rather a familiar 'ugly' parade of highly localised degradation is offered. Nevertheless, there may be merit in requiring miners to rectify disturbed land.

One approach is to demand the posting of environmental bonds; however it may be more efficient to ensure appropriate rectification through the agency of insurance companies (insurance could, of course, evolve side by side with the posting of bonds). Swedish law has accepted insurance mechanisms against inadvertent harm as an alternative to exacting environmental charges. Where mining in an area might lead to quite dramatic changes, and the area itself is considered to have valuable recreational or bequest values, measures which allow sequential development might be considered. These could include auctioning rights or trading pre-existing rights. The transaction costs of such measures may prove to be insuperable but are worth investigation as alternative approaches to the total denial of access in national parks.

Mineral extraction, by definition, involves transforming the pre-existing resource. In the main, however, mining is confined to small and isolated locations. Typically, according to evidence supplied by CRA to the Industry Commission, a successful mining operation will occupy less than one square kilometre within an original exploration lease of 1000 square kilometres. Moreover, as the naturalist Harry Butler has made clear (ABC,1990), almost all land which has been mined can be readily restored. In the case of beaches, restoration is particularly straightforward; indeed operators of tourist buses on the northern NSW coast ask their clients to nominate which of the beaches they see before them have been sand mined, knowing that the differences are imperceptible.

For other areas, restoration may be more costly and involve retaining different layers of the original soil and replacing it. Some estimates from open cut mining operations suggest that costs of thorough rehabilitation could range into the billions of dollars. Even so, such procedures are now widely practiced. Doubtless this adds costs to mine development—costs which in restoring the land to its original condition might be several hundred fold the value of the restored land. Quite clearly, for land worth only thousands of dollars subsequent to its rehabilitation, society would obtain poor value for the money spent were we to insist upon rehabilitation.

Accordingly, decisions on rehabilitation might best be left to prior agreement between the miner and the landowner (who may be the government); with a court establishing the measures to be taken, or compensation to be paid, in the event of disagreement. The basis of such decisions would need to be specified. Decisions would, for example, have to be closely related to the value of the land in close proximity to that which has been disturbed. Such approaches are far more effective in internalising externalities than those requiring specific actions to be taken

For these reasons, whilst it may not seem unreasonable to expect rectification of land after mining has ceased, requiring such costs to be absorbed may be irrational in both an economic and a wider social sense. There may be room for bargaining solutions to be adopted based on the value of the land being returned to its original condition and the cost of undertaking this activity.

Concluding comments

Mining contributes appreciably to the generation of wealth and the material satisfaction it brings must be traded of where its impact adversely affects unpriced values like wilderness. Moreover, in such trade-offs, we should be mindful that increased wealth allows us to afford things otherwise unavailable to us. Placing obstacles in the way of the wealth generation which mining development brings is more likely to impede the pursuit of overall community well-being than to prevent destruction of natural assets which would be valued greatly now or in the future. A key to the appropriate policy approach is to recognise limitations on the availability of information. Policy must ensure the correct structuring of incentives to enable information to be economically generated and applied—information on needs of markets and the ways to meet these needs. Equally important, policy must ensure measures do not impede this information discovery process, or the uncovering of hidden values.

The essential goal of societal organisation for mining, as for other activities, is to set a framework within which people can pursue their ever changing aims on the basis of their own knowledge.

It is apposite to quote Hayek on competition in setting out the preferred regime under which mining should operate. Hayek put his view in the following way:

> Competition is..., like experimentation in science, first and foremost a discovery procedure. It is irrelevant... (to)...

evaluate the results of competition ... from the assumption that all the relevant facts are known to some single mind. The real issue is how we can best assist the optimum utilisation of the knowledge, skills and opportunities to acquire knowledge, that are dispersed among hundreds of thousands of people, but given to nobody in their entirety. Competition must be seen as a process in which people acquire and communicate knowledge; to treat it as if all this knowledge were available to any one person at the outset is to make nonsense of it. And it is as nonsensical to judge the concrete results of competition by some preconception of the products it 'ought' to bring forth as it would be to judge the results of scientific experimentation by their correspondence with what had been expected. (Hayek, 1976: 67).

It is competition that reveals the discoveries on which wealth and well-being generally are founded. Institutional arrangements for mining are best focussed on allowing this process to unfold rather than seeking to extract shares from the income it generates. It is even more unwise for government instrumentalities to attempt to determine in advance how the process of income generation should unfold. Mineral resources are effectively undepletable; and we should not, therefore, establish frameworks which are appropriate only for sharing a scarce resource which is in fixed supply. Whilst farming, the tall chimney stacks of heavy industry and the exhaust pipes of motor vehicles all make extensive use of common environmental goods; workable mineral deposits are located as pinpricks on the map. The extraction of their value imposes relatively slight damage; but where that damage might be appreciable, market-based mechanisms, involving appropriate charges, should be established to ensure adequate compensation to the community and incentives for producers.

7 Markets and sewerage

Executive summary

The sewerage industry in Australia employs assets with a written down replacement value of around $18 billion (1987–88), which compares with $13 billion for Telecom and $1.5 billion for Australia Post (Industry Commission, 1990). Its task is to transport domestic and industrial liquid wastes away from cities and towns to sites where the environmental impacts of waste discharges can be managed. As a part of this task, the industry may treat the sewage, both to reduce its disposal costs and to reclaim valuable constituents for sale.

Once liquid wastes are securely enclosed in pipes and access to the sewer is controlled, sewer disposal of liquid waste is exclusive to the purchaser, and can be bought and sold like any other marketed service.

Liquid wastes, as by-products of both domestic and industrial activities, are collective 'bads' once they enter the environment, in the sense that many people are affected by the same waste discharge. Due to large numbers of people affected, and a lack of defined individual property rights to a clean environment, market bargaining will not signal damage costs back to sewer operators and the generators of domestic and industrial wastes. In practice, environmental damages are assessed, and permitted levels of discharges of liquid wastes from sewers and other sources determined, by state environmental protection agencies.

Where waste discharges are restricted, rights to dispose of environmentally damaging liquid wastes via the sewers or otherwise are scarce goods. In the absence of market signals to indicate the value of scarce liquid waste disposal rights, different parties along the sewer pipe will make decisions in ignorance of the benefits and costs of their choices for others.

As Australian sewerage is currently organised, individuals and firms have little incentive to consider the effects of their liquid waste production and disposal on others. There are few market price signals to indicate the value of rights to sewer disposal of liquid wastes. The managers of publicly owned and operated water and sewerage enterprises have not been encouraged to price according to the costs of providing services. Sewage and its constituent wastes are rarely measured, and quantity-based charges for sewerage services and for discharge into the environment are the exception rather than the rule. Without such market signalling and incentives, Australians do not get the best possible mix of waste production, transport and treatment for permitted discharges into the environment.

Market-based approaches to liquid waste disposal and sewerage have been adopted in several countries overseas. Privatisation of sewerage enterprises, recently undertaken in the UK, and franchising of sewerage operations, practised in France, improve the operational efficiency of sewerage enterprises, but create little new information or incentives for waste producers and processors to coordinate their activities. Effluent charges, as adopted in Germany, do provide information and incentives for parties to recognise and respond to their impacts on others. However, since effluent charges set prices and not quantities of discharges, they are unattractive to control agencies who prefer control over quantities when they cannot accurately measure pollution damages.

Marketable liquid waste discharge rights, although yet to be adopted anywhere for sewage discharges, appear the most promising of the new institutional innovations. Marketable rights equal to the present permitted levels of discharges accomodate the present quantity standards approach to pollution control. New rights could be created when sewage disposal facilities are expanded. The prospect of profits from market trading in rights provides both the incentive to reveal information about benefits and costs of waste disposal and the incentive for parties to respond to information provided by others. The values of rights will also provide information about the need to expand the system. Discharge rights will end up with those who value them most, and thus the overall cost of achieving permitted discharge standards will be minimised. The fact that marketable rights would be considered private property will be a desirable constraint on ill-informed or politically biased alterations to environmental standards.

The present high costs of accurate and reliable liquid waste measuring apparatus are a major barrier to the prompt introduction of marketable rights for liquid wastes, since the value of rights is dependent on careful policing of all discharges, to ensure that individual contributions do not exceed entitlements. Effective marketable rights to sewer disposal of wastes will require accurate measurement of the particular waste at all sewer intake and exit points. The same accuracy of measurement will also be required for implementation of effluent charges, if they are to achieve a similar degree of efficiency in liquid waste disposal. Both of these market-based approaches to environmental protection bring the high costs of effective and efficient environmental protection out into the open, mainly in the form of high measurement and policing costs. The current 'command and control' approach, involving modest scientific research budgets, little accurate measurement, and little incentive to honestly reveal individual costs and benefits, tends to conceal the high costs borne by individuals and industries under current quantity standards.

Creating markets in sewage wastes will be costly and will take time to implement, but, over time, the institutional change will itself help to solve the technical problems. The past organisation of sewerage and environmental protection has discouraged research and development in the areas of liquid waste treatment and measurement; the prospect of markets in liquid waste discharge rights would stimulate efficient research, investment and innovation in all these areas.

Introduction

An important and valued achievement of modern industrial society has been the divorce of household and industrial liquid waste production from the need to worry about its disposal. We not only create liquid wastes directly; we also cause industry to produce wastes in the process of satisfying our needs. By purchasing commodities and services we require someone to dispose of the residual wastes from the productive process. Similarly, we pay fees or local taxes for the disposal of our own personal wastes. We pay little or no attention to the processes involved in these activities, legitimately insofar as our payments for the products and services are assumed to acquit our personal reponsibilities for the costs of liquid waste disposal. Thus it comes as rather a shock if disposal is unsuccessful, as when we encounter sewage effluent in the waves at Bondi (Beder, 1990).

There is nothing wrong with people not understanding all the implications of their day-to-day decisions for liquid waste production and disposal. The economy and the environment are complex, and time is precious. A great advantage of markets is in allowing economies in the search for information. Those with specialist knowledge, skills and capital are contracted implicitly or explicitly to provide the necessary information about the costs and benefits of waste production and disposal. What is required is the correct signalling of costs, so that when individuals or firms make decisions which lead to liquid wastes and waste disposal, they consider the costs imposed on others as well as themselves. If they do, the community should get about the right amounts of liquid waste generation and disposal—the amounts which correctly balance the benefits of waste-generating domestic and industrial activities against the costs of waste transport and treatment and the unpleasantness and harms due to waste discharge into the environment.

As sewerage is organised in Australia, do individuals and firms in fact have the necessary incentives to consider the effects of their liquid waste production and disposal on others? We conclude that the present publicly-owned and operated sewerage systems blunt such incentives, and outline alternative methods of organising sewerage which would provide more market signals to prompt people to change their behaviour in the interests of others. Problems involved in implementing market-based organisation of sewerage are also examined. But first we briefly consider the nature of sewage and the general relationships between discharges and impacts on the environment.

Wastes in sewage and environmental impacts

Sewage includes human wastes which are in effect liquid, plus wastes which can be conveniently transported in water in solution or suspension, such as soil particles, food particles and grease. Sewage is thus a complex mixture of mineral and organic matter, in suspension or dissolved, comprising 99.9% or more of water (Bolton and Klein, 1971).

The flow of water in a sewer system provides both waste transport and dilution, a role which people have assigned to rivers throughout history. Sewers are artificial, enclosed rivers specifically created for waste transport; they cannot function without a continuous and substantial water supply to waste-producing households and firms.

The common constituents of sewage include:

- suspended inorganic solids, such as sand and clay;
- oxygen-demanding organic wastes, normally measured by their biochemical oxygen demand (BOD);
- nutrients, such as nitrogen and phosphorus;
- pathogens, such as bacteria and viruses;
- oil and grease;
- inorganic synthetic chemicals, such as detergents, and
- heavy metals, such as lead and mercury.

Each of these and many other sewage constituents has an impact on the environment where it is discharged, and particular constituents may interact in the sewerage pipe or in the receiving environment. Thus monitoring of sewage discharges aimed at avoiding environmental impacts harmful to people may require many separate measurements, and estimates of environmental impacts are subject to considerable uncertainty.

When treated or untreated sewage is discharged into the environment, the wastes it contains impose costs in the form of unpleasantness or harm on people who use that environment. To understand the relationship between discharges and costs to people, recall that the natural environment has a capacity to absorb most wastes, either by transforming them into harmless substances, or by diluting or dispersing them so that the resulting concentrations are harmless (Tietenberg, 1988). The relationship

Chart 7.1

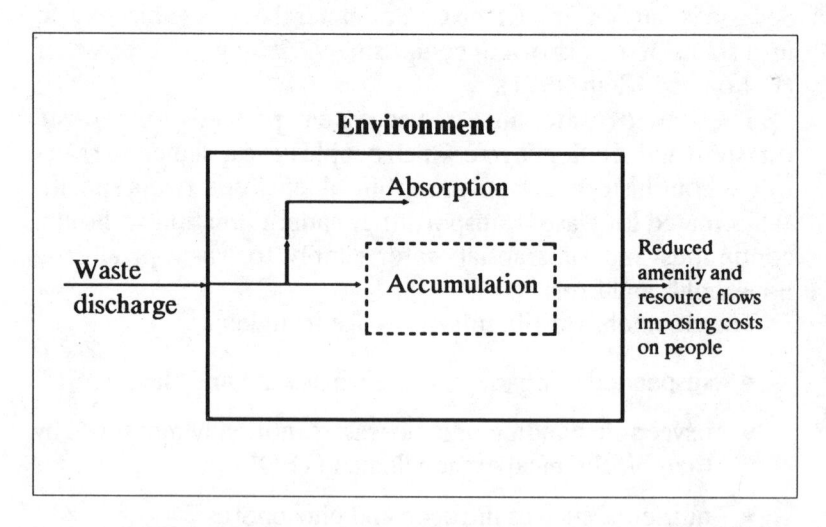

```
                    Environment
                      Absorption
                    ┌─────────→
                    │        ┌──────────────┐      Reduced
    Waste           │        ┊              ┊      amenity and
    ─────────────→──┴──────→ ┊ Accumulation ┊ ───→ resource flows
    discharge                ┊              ┊      imposing costs
                             └──────────────┘      on people
```

between the rate of waste discharge and the rate of environmental absorption is critical in determining the nature of changes in the natural environment and consequent costs to people. This relationship is illustrated in Chart 7.1.

Chart 7.1 suggests three main types of relationships between discharges and environmental change. The rate of discharge of a waste may be less than the rate of absorption. Second, the rate of discharge may exceed the rate of absorption, so that the waste accumulates, but the environment can recover, that is, environmental absorption of the waste can reduce the accumulated stock if discharges are reduced or cease. Finally, absorption may be non-existent or negligible, so discharges accumulate over time, and environmental change is permanent.

For example, consider sewage containing organic matter and heavy metals discharged into a stream. So long as discharges of organic matter are less than absorption, there will be a temporary but not drastic change in the environment, in the form of a lowering of the dissolved oxygen content of the stream. If organic discharges exceed absorption, organic wastes accumulate in the stream, which becomes putrid when all its dissolved oxygen is used up. The stream will begin to recover once discharges fall below absorption. In the case of the heavy metals, the wastes accumulate continuously, and effects on the environment and consequent harm to people increase for as long as the discharges continue.

The natural rate of absorption of biodegradable wastes will vary enormously depending on the volume, temperature, chemical composition, mobility, rate of oxygenation, and so on, of receiving waters. Rates of absorption in confined, immobile inland waters will generally be low; on the other hand, scientific studies commonly record very little environmental impact more than 100 metres or so from ocean outfalls releasing large volumes of partially treated sewage (Holmes, 1987).

The implications for environmental protection measures are clear. Where absorption is substantial, rates of waste discharge must be the object of control, with the aim normally being one of keeping discharges below the absorption capacity. Short-term surges in discharges, causing temporary waste accumulations, will have to be monitored separately, in order to compare incremental benefits and costs. Where absorption is negligible, the accumulated stock of the waste in the environment must be monitored, and the resulting permanent costs assessed, as a basis for comparing the incremental benefits versus (continuing) incremental costs of adding to the stock.

Some economics of liquid wastes

Liquid wastes as collective bads

Due to their unpleasant or harmful nature, liquid wastes are 'bads' to the people who produce them as by-products of domestic or commercial activities, that is, the producers are willing to make sacrifices to be rid of the wastes. The cheapest form of disposal for settled individuals, generally practised by people until late last century, is to dump the wastes just far enough away to avoid the nastiness (Butlin, 1976: 9–11; Fogarty, this volume). This does no harm in the absence of neighbours, but others are harmed when people congregate in villages or towns or cities. Such costs, imposed on others without their consent, are termed externalities or spillovers.

Spillovers between neighbours can be overcome if the waste disposers can be forced to bear the full costs of the smells, health problems, damage to plants and animals, etc., that they impose on others. However it will not always pay to do so; sometimes the costs of private contracting or court action or government enforcement of pollution controls will exceed the costs of the spillovers. The community will be better off if no formal action is taken. An individual may protest to his neighbour if the neighbour's dog relieves itself on the first person's lawn; taking the neighbour to court would involve disproportionate costs for all concerned. As in many other aspects of human interaction, informal restaints resulting from common courtesies and mutual sensibilities are often more effective than formal legal constraints.

Dispersal of liquid wastes in the environment, via water or air movements, and the mobility of people, mean that liquid waste discharges commonly affect substantial numbers of people. The wastes are collective or public bads, in the sense that many people are harmed by the same waste discharge. In these circumstances, dischargers are unlikely to bear the full costs they impose on those harmed by waste discharges. There are two main reasons for this. One is that it is often difficult to identify who discharged the offending waste, so that market deals or legal action to correct the problem are impossibly expensive (*The Sunday Herald*, August 12, 1990).

The second reason, why dischargers are unlikely to bear the full costs they impose on others, is that individuals harmed by discharges often have little incentive to act to signal their costs back to dischargers. To see why, consider the case of an individual fisherman or riverbank resident downstream from an unregulated pulp mill. Knowing that he is one of many, the individual who is harmed by discharges from the mill may prefer to free-ride—to leave it to his fellow fishermen and residents to pay and/or lobby to have discharges reduced. After all, his contribution will make little or no difference, and he will reap the full benefit of any reductions. Thus, quite rationally, he may decide to make no contribution at all. If many or most affected parties downstream think likewise, little of the total costs of liquid waste disposal is imposed on dischargers, and the community suffers from excessive disposal of liquid wastes.

Sewerage—sheeting home the costs

The collective bad nature of liquid wastes, and the consequent failure adequately to control private waste disposal, caused major public health problems in urban areas up until the late nineteenth century. Publicly-funded and -operated sewers were created to segregate liquid wastes from the urban environment and to provide water transport of wastes to sites where their impact on the environment would be less costly and harmful.

Once liquid wastes are securely enclosed in pipes, access to the sewer system may be controlled and other methods of waste disposal strictly policed and penalised, the waste-producing household or firm cannot shift the costs of waste disposal onto others. The sewer is the least-cost method of disposal, and the waste producer can be forced to pay the cost of sewerage by the threat of exclusion, which would expose the household or firm to even higher anticipated costs of on-site treatment or storage, illegal dumping penalties or shutdown.

Strict policing of sewer access and illegal dumping of liquid wastes creates a willingness to pay for sewerage disposal of wastes, stimulating the provision of sewerage services. At the same time, sewerage charges may encourage households and industries to reduce their production of liquid wastes, and may create incentives for treatment of wastes to render them suitable for disposal on-site.

Households and firms are legally liable for the disposition of the wastes they produce while the wastes remain at the production site. Once in the sewer, the wastes from separate sites are rarely identifiable. Legal responsibility for the disposition of wastes in the pipe resides with the sewer operator, who should be subject to penalties for unauthorised discharges which harm others.

At the sewer outlet, large volumes of sewage are dispersed into coastal or enclosed waters, imposing costs on large numbers of users of that environment. As in the case of local disposal by waste producers, private action will not accurately reflect affected parties' costs back to the sewer operator, even if the collective costs to those parties are large. Individuals who are harmed—fishermen, bathers, birdwatchers, coastal residents—don't each own a piece of the receiving waters, and thereby have the right to charge the sewer operator for disposal, or sue for damage to their property. Knowledge that the number of parties harmed is large inclines each individual to leave it to others to take action. Also any negotiations between individuals and the sewer operator over damage compensation or payments for reduced discharges would be complicated by the difficulty of assessing affected parties' true costs.

In practice, environmental damage around sewer outlets is assessed, and constraints on sewage discharges determined, by government agencies. Normally a state government environment protection or pollution control agency determines allowable discharges of particular wastes from all sources, and public and private sewer operators are issued with permits which authorise discharges up to those limits.

To get discharge standards right, an environmental protection agency needs to know how the costs of sewage discharges to all those harmed, and the benefits of discharges to households and firms, vary as the level of discharges varies. Ideally, the agency would permit that level of discharge at which the total cost to affected parties of the last unit of sewage discharged just equalled the benefit (profit) to households (firms) of being able to discharge that unit of sewage (see Chart 4.1 above, with levels of sewage discharges replacing levels of land degradation). But because sewage discharges are collective bads, there is still a problem in identifying the true costs of those who lose due to discharges. Individuals who are harmed have little incentive to reveal their true costs to the agency; on the contrary, they have incentives to exaggerate costs, to persuade the agency to set tight standards, so long as they do not have to pay for the standards. So the introduction of a government agency, with the power to set discharge standards and to force compliance by the sewer operator, does not guarantee the 'right' amount of sewage discharge, even if we ignore the complex constitution of sewage and changing conditions in the receiving environment.

The demand and supply of sewer waste disposal

Where disposals of wastes are restricted, rights to dispose of environmentally damaging liquid wastes via the sewer are scarce goods. The rights are valuable to households and firms because the legal alternatives to sewer disposal of wastes, on-site treatment and cessation of waste-producing activities, are all costly. The value of such rights depends on the supply of and demand for sewer disposal of particular types of liquid wastes.

The supply of sewer disposal of particular wastes is determined by:

- permitted discharge levels for each type of waste in the receiving environment;
- the costs of constructing and operating the sewer system, and
- the costs of constructing and operating any sewage treatment plants which remove wastes from the system.

The demand for sewer disposal of particular wastes is determined by:

- households' willingness to pay for liquid waste disposal;
- the willingness of commercial and industrial firms to pay for different levels of waste disposal, which is bounded by the profits made by firms at different levels of waste discharges, corresponding to different levels of output and resource use, and often to different production technologies, and
- the costs of alternative methods of liquid waste disposal, in particular, of processing to reduce liquid wastes, net of the values of valuable materials recovered.

These separate determinants of supply and demand are mostly under the control of separate actors—environment protection agencies, sewer system operators, treatment plant operators, households, commercial and industrial firms, liquid waste processors and potential entrants to the industries involved. For efficient provision of liquid waste disposal, it is essential to have accurate communication between these parties about waste producers' disposal needs and willingness to pay, permitted discharge levels, and the disposal alternatives available and their costs.

Monopoly in sewerage

Sewerage providers are rarely subject to competition or the threat of competition in the transport of sewage, due to the very high costs of duplicating pipe networks. Since households' and firms' alternative means of liquid waste disposal are usually much more costly, the sewer system operator is able to raise charges above the true costs of providing the service. Hence sewerage is invariably subject to price regulation, either via public operation of sewer systems, or direct regulation of firms' prices.

The existence of co-tenancy contracts for co-ownership of large indivisible facilities, such as generators and transmission lines, in the Unites States' electricity industry suggests a possible avenue of escape from the pipeline monopoly problem (Smith, 1988). This would involve joint ownership of pipes, with capacity rights being proportional to shares of total asset values. Capacity rights in pipes would be marketable, and each co-owner would pay usage charges for measured use of the pipes. A pipeline would be managed by a separate operating company created by the co-owners. Thus it may be possible to introduce competition between water and sewerage providers who use the same pipes.

The market power of sewer operators will be much less where waste producers have low cost alternatives to sewer disposal of wastes. This will often be true for the producers of small volume industrial wastes, where road transport and other forms of disposal such as incineration are available, and for large firms and industries able to arrange their own disposal outside the sewer system. For example, Victoria's State Electricity Commission has constructed a separate pipeline to transport saline wastewater from its Latrobe Valley power stations to Bass Strait.

Policing and measurement of liquid wastes

Policing makes waste producers liable for the costs of the wastes they create; they can't shift the costs onto others. Policing plus penalties create incentive to reduce waste production and willingness to pay for waste disposal and treatment. With the prospect of reward for their services, individuals and firms have incentives to establish liquid waste transport and processing businesses.

The need for policing is unaffected by the mechanism

employed to reflect costs back to waste producers. In principle, this might be achieved either by creating private property rights in an unchanged environment, allowing individual victims to impose costs on waste dischargers, or by government charges or penalties imposed on behalf of victims.

Unless the object of policing is to eliminate waste discharges into the environment, policing must involve measurement of liquid wastes. Measurement of the contributions of individual liquid waste dischargers is necessary to establish their liability for costs imposed on others. For example, an environmental protection agency needs accurate discharge data to impose penalties on firms violating their discharge permits, and the court will require the same in a compensation case brought by victims of discharges.

Measurement of the amounts of liquid wastes transported and processed is also desirable to establish the true costs of disposal to waste producers, and the appropriate rewards due to parties who perform these valuable activities. Sewage is commonly unmetered, with disposal charged on a lump-sum basis. This may be justified by high costs of measurement of sewage volumes and constituents. However, as a result, the sewer operator has no means of setting quantity-based charges which encourage lower discharge levels. It is unable to communicate information about the costs which extra quantities of wastes from households and firms impose on itself and on the environment at the sewer outlet.

The fact that sewerage is for the most part a monopoly service reduces the pressure for accurate measurement of liquid wastes. With no alternative suppliers of sewage disposal breathing down its neck, a sewer operator is in a strong position to unilaterally revise contractual terms. On the other hand, liquid waste disposal firms operating in a competitive environment will want to measure the quantities of various wastes to ensure that the costs of their services are covered. Waste producers contracting with commercial waste disposal firms will also require measurement to ensure that they receive the services they contracted for. Thus striking a deal will involve measuring the quantities of the particular wastes involved.

Since victims of liquid waste disposal may suffer damages based on either the rate of discharge or the accumulated stock of particular wastes in the environment, it is necessary to identify the correct damage measure for each type of liquid waste.

Organisation of sewerage in Australia

Industry organisation

In Australia, sewerage is everywhere publicly provided. Where both water and sewerage are provided, they are almost always combined in a single enterprise. In the major cities, sewerage is provided by statutory authorities or city councils. Elsewhere, services are provided by local councils or regional boards. Historically the operations of sewerage enterprises in Australia have always been open to political influence; appointments to water and sewerage boards have typically reflected political jurisdictions and other interest groups as well as technical and business expertise.

Water and sewerage enterprises in Australia are currently responsible for the provision of a variety of commercial and noncommercial services, ranging from water supply, sewerage and drainage control, to regulation of industrial waste discharges and land development and management of parks and wildlife. Almost all of their income is derived from the provision of water and sewerage. They are not explicitly required to earn a competitive rate of return on assets. Consequently there has been inadequate incentive for management to upset traditional modes of operation by moving to reduce costs or to set prices on the basis of the costs of providing services. For example, water and sewerage enterprises generally undertake a high proportion of construction and operating activities in-house, rather than contracting jobs out to specialist enterprises. Moreover, because the charges for the different services provided are typically unrelated to costs, cross-subsidies between different services, and between different customer groups, are common. Thus in 1987–88, the Sydney Water Board's domestic customers were subsidised in their usage of water and sewerage services by 67,000 industrial users to the tune of $140 million (Australian Financial Review, February 16, 1990).

Discharges of liquid waste are typically not metered, and

sewerage charges are generally based on property values and unrelated to quantities of wastes removed. Volume and concentration-based charges for industrial wastes are progressively being introduced. Charges are commonly based on several broad categories of pollutants including suspended solids, biochemical oxygen demand, nitrogen, sulphur, and phosphorous (for example, see Melbourne and Metropolitan Board of Works, 1990). Quantities of these and other pollutants discharged into the sewers are limited by trade waste agreements which are established between industries and the sewerage authority in question. Where liquid waste produced by industries is not suitable for discharge into the sewer, private firms provide liquid waste collection, treatment, recycling and disposal services for industrial wastes.

State environmental protection agencies issue licences specifying maximum allowable quantities and concentrations of discharges of liquid wastes. Liquid waste producers can negotiate with the agencies about direct discharges into the environment. Sewerage enterprises are similarly required to obtain licences for their discharges. Monitoring of compliance is the responsibility of the agencies, which have the power to prosecute dischargers which fail to meet their licence conditions.

The effectiveness of the policing of these standards is unclear, as resources available for monitoring illegal discharge are limited, and the ability to trace pollutants to their source is constrained by the current state of technology. Furthermore, the determination of environmental standards is not transparent; it involves a mix of scientific evidence, the current use of the receiving environment, environmental standards elsewhere, advisory bodies of experts and interest group representatives and political lobbying (see Bates, 1983: 157–164).

Efficiency consequences of the present organisation

The organisation of sewerage affects the efficiency of resource use at two levels: within sewerage enterprises, where efficiency depends on managers having appropriate incentives and on signalling and incentives within the enterprise, and in allocation of resources along the length of the sewer pipe, where the separate actors require appropriate signals and incentives to communicate information about their waste disposal needs and alternative disposal options.

The average real rate of return for Australian water and sewerage enterprises was approximately 2% in 1987–88; the rate for metropolitan sewerage was 3.2% (Industry Commission, 1990). This low level of return is a reflection of the organisational structure and accountability arrangements for these enterprises. Board appointments are made for political as well as commercial reasons. Water and sewerage enterprises are commonly required to achieve environmental and public interest objectives, some of which are very difficult to assess, as well as commercial objectives. Boards are subject to Ministerial direction. As a result, managements produce services in an inefficient manner (for example, by allowing overmanning which satisfies management preferences for a 'quiet life'), and may fail to produce services desired by people (for example, slow response to service complaints).

A correct allocation of resources along the sewer pipe depends partly on setting the correct discharge standards at the sewer outlet. However, in responding to a mix of scientific evidence, expert opinion and political pressures, Australian environmental protection agencies are unlikely to get discharge standards right. As explained above, what they really need to know are the costs of particular waste discharges to victims and the benefits of the same discharges to households and firms. The costs to victims are obscure due to the 'collective bad' problem. On the other side, if households and firms are not charged for their waste discharges, the environmental protection agency has no direct information about the benefits they receive. This is the situation for most sewerage users, as distinct from industries producing certain liquid wastes.

In the absence of measurement and market-based charges for sewer disposal of liquid wastes, there is little accurate communication along the sewer pipe about what current and potential waste producers—households, firms, liquid waste processors—are willing to pay for sewer disposal, and what those who supply disposal facilities—sewer system operators, treatment plant operators, environmental protection agencies who control access to the environment—are willing to accept in compensation. Yet when a household or firm considers increasing its output of liquid wastes—installing a dishwasher or an additional killing chain at an abattoir—it should have information about, and the incentive to allow for, the costs of liquid waste transport and treatment and any resulting environmental damage which its decision imposes on others. Similarly, a sewer or treatment plant operator considering a sewer extension or additional treatment capacity requires information about the consequent benefits to waste producers, and revenue derived from service charges will normally provide the incentive to undertake the project.

Market-based organisation of sewerage

Governments in a number of OECD countries have moved from traditional 'command and control' mechanisms of regulation of liquid waste disposal to more market-oriented arrangements designed to improve information signalling and incentives. Some of these organisational reforms, such as the recent privatisation of water and sewerage authorities in the UK and franchising of water and sewerage operations in France, have been designed to improve the operational efficiency of sewerage enterprises. Others, such as the imposition of liquid effluent charges or taxes in some European countries, and proposals for the creation of marketable liquid waste discharge rights or permits, so far only implemented in isolated cases (OECD, 1989: 95–97), are designed to improve the allocation of resources between parties along the sewer pipe.

Privatisation/franchising of sewerage

In the face of budgetary constraints, aging facilities, and internal and external pressures to upgrade treatment requirements, some countries have divested the management and, in some cases, the ownership, of sewerage facilities from the public sector, relying more heavily on private provision of sewerage services. In the UK, the responsibilities of ten water and sewerage authorities were transferred into private hands in December, 1989, when the authorities were transformed into public limited companies (PLC's) (Kinnersley, 1988). Privatisation has been coupled with a strict regulatory regime. Regulation of prices and service quality is seen as a necessary constraint on the exploitation of monopoly power by the PLC's, which are the sole providers of sewerage services in their respective areas of operation.

The UK privatisations have improved sewerage managers' incentives to minimise costs and to provide services desired by sewerage customers. However, since there is only limited measurement of liquid wastes, the system does little to improve signalling of the costs and benefits of waste discharges.

Privatisation of the management of sewerage facilities has also been occurring in France and the US. In the US, where the burden of constructing and maintaining sewerage facilities falls on local councils, augmenting and replacing the present infrastructure is proving a substantial drain on government budgets. Local councils are seeking out savings by contracting with private firms to operate sewerage facilities (Stump, 1986). Many French municipalities contract the management of water and sewerage facilities to one of five private water and sewerage companies which operate nationwide. About 40% of sewage disposal in France is managed by the private companies (Deschamps, 1986). Typically, the municipalities maintain ownership of the basic infrastructure, and franchise management of the facilities to private companies (Kinnersley, 1988). Competitive bidding for franchises encourages the firms to provide acceptable quality services at low cost. Municipalities retain the right to resume management of facilities at any stage. The major problem with the franchise contracts is specification of the terms of maintenance of water and sewerage assets which outlast the term of the franchise.

Liquid effluent charges

Governments are also increasing their use of effluent charges in pollution control, including charges for sewer disposal of wastes (OECD, 1989). Most OECD countries impose some charges on liquid wastes discharged via sewer systems. Some OECD countries, including Australia, also allow direct industrial discharge into the environment. Firms are required to comply with licences issued by the relevant government agencies, and pay a volume- or concentration-based charge for the right to dispose of their effluent in this manner.

OECD countries have been reluctant to rely solely on effluent charges to control the discharge of liquid waste into the environment (Hahn, 1989). Since the 1970s, combined systems of regulation plus charges have been adopted in a number of countries, including Germany, France, Italy and the Netherlands. Effluent charges based on the estimated value of environmental damage caused by liquid waste discharges would signal the costs of liquid wastes to producers, thereby creating incentives for waste treatment and the reduction of discharges by waste producers. Most actual effluent charges understate the costs imposed on the environment by liquid waste discharges; the charges typically function as revenue-raising rather than cost-signalling devices (Hahn, 1989).

The effluent charge system in the Federal Republic of Germany has an explicit incentive objective, designed to encourage firms to meet licence specifications. The important feature of the German charge system is variable charges, based on the degree of compliance with environmental standards. If federal minimum effluent standards are met, as specified in industrial permits, a discount of 50% is applied. Where industrial discharges fall below 75% of minimum standards, firms are charged according to the actual volume of liquid waste discharged. Payment of the charge may be delayed for three years where firms adopt waste treatment technology capable of reducing industrial discharge by 20% (OECD, 1989: 39).

Marketable liquid waste discharge rights

Marketable rights (also commonly termed marketable permits) fit neatly with the almost invariable preference of environmental protection agencies to set quantities rather than prices of liquid waste discharges (OECD, 1989: 111–112). If discharge rights are, like any other private property, exclusive to their owner and transferable in the market, they will eventually end up with those who value them most. The market will reallocate rights over time so that the community gets the greatest possible returns net of costs of transport and treatment of wastes from the permitted maximum level of pollution.

Marketable rights for liquid waste discharges have only been implemented in a handful of cases worldwide (OECD, 1989). They have not been introduced for sewage waste discharges. Yet they are straightforward enough in concept. Liquid waste discharge rights could be created for each category of discharge into the environment from a given sewer outlet. Rights would entitle the owner to discharge specified quantities of wastes into a specified sewer system. Rights equal to the total permitted discharges from each sewer outlet would be allocated initially on some basis for each category of waste. The rights would be tenable for a long period and tradeable free of controls. In principle, rights to discharge domestic and service industry wastes could be issued to individual households and small businesses; in practice, as explained below, waste measurement costs and other costs of issuing and trading rights will make this prohibitively expensive for such small liquid waste producers.

For each liquid waste category, parties with an interest in trading in rights will include the original waste producers, waste treatment firms which use the sewers, sewage treatment plants and the sewer system operator. Individuals and groups harmed by pollution, and the government, could enter the market to buy permits and retire them, thus reducing total permitted discharges.

Implementation of market trading in rights creates incentives for would-be buyers and sellers of rights to reveal the values they attach to rights to discharge liquid wastes. It thereby encourages communication of benefits and costs of liquid waste production, transport and treatment between parties along the sewer pipe. This will promote efficiency in production, transport, and treatment of liquid wastes. Compared with the current non-tradeable discharge licenses, highly-profitable firms with no cheap way of reducing liquid waste production have the additional option of purchasing discharge permits from other firms. Less profitable firms, and firms that can reduce their discharges cheaply, have the extra option of selling permits at the same time as they change products or production methods. As a result, over time the overall cost of achieving permitted discharge standards will be minimised.

The industrial chemical complexes around Altona in Melbourne and Botany in Sydney typify situations where ability to trade liquid waste discharge rights might be mutually beneficial to participating firms. Where two firms produce the same liquid waste, but face different costs of waste reduction or treatment or different market prospects for their products, trading may increase the expected profits of both firms. Trading in discharge rights may also provide a 'golden handshake' for firms in declining industries; for example, the closure of textile processing firms which are major polluters and only marginally profitable may be facilitated by sales of such rights.

Charges versus marketable rights

Economists have debated extensively the theoretical and practical merits of effluent charges versus marketable rights. By most accounts, the correct choice depends on circumstances (Baumol and Oates, 1988). Charges set the price of liquid waste disposal; rights set allowed quantities. Each put a price on scarce rights to sewer disposal of environmentally damaging liquid wastes; charges directly, by government decision, and marketable rights indirectly, by forcing would-be dischargers to compete in the market for the limited supply of rights. Each thus prompt interested parties along the sewer pipe to reveal information about willingness to pay for waste disposal and the costs of alternative methods of disposal.

Charges and marketable rights generally have different effects on the incomes of waste producers, dischargers and the wider community. Charges automatically embody the polluter-pays principle, and cause a redistribution of income away from liquid waste producers and dischargers to government and thence to the wider community. Government auctions of marketable rights can in principle achieve similar distributional outcomes to charges; in practice the rights are invariably given to the holders of current discharge licenses. The effluent charges actually imposed in European countries are generally well below estimated environmental damage costs, and charge revenues are commonly used to subsidise private investment by industry in pollution control. These facts suggest that a high weight is given to the status quo distribution of income in political decisions about the environment (Hahn, 1989).

Another important difference between charges and

marketable rights is the ongoing role of government. Charges remain under government control, and are able to be changed in response to new scientific, economic or political information. Once established, rights will become jealously-guarded private property, a basis for longer-term decision making, and correspondingly more difficult to change. As knowledge of the environment and desired discharge standards change, those with a benign view of government will see charges as facilitating desirable flexibility in environmental policy. Marketable rights, probably requiring government purchase of rights or compensation when standards are tightened, will be seen by others as a desirable constraint on poorly-informed or self-seeking governments.

Markets in liquid waste discharge rights

Since Australia's environmental protection agencies rely on quantity standards, rather than charges, market signalling of the costs and benefits of liquid wastes would be most conveniently introduced by creating marketable rights to sewer disposal of liquid wastes.

Advantages of marketable rights

As previously explained, marketable rights create the incentive to reveal information about the costs and benefits of waste production, transport and treatment, and thereby help the community to achieve effluent discharge standards at least cost. So long as there are no restrictions on who can buy rights, the incentive extends beyond current players; if existing companies (in particular, international waste treatment companies) or new companies using Australian scientific and technical expertise wish to enter waste-producing industries or waste treatment, they can do so by purchasing rights. This will encourage the adoption of new technologies in both the waste producing and waste treatment industries.

Measurement problems

If the creation of marketable rights to the sewer disposal of liquid wastes would save a lot of community resources, why has no market economy implemented it? First and foremost because of the complex composition of sewage and the high costs of accurate measurement of most liquid wastes. Recall that sewage commonly contains a large number of wastes with distinct effects on the environment which impose costs on people. Separate marketable rights are required for each type of harmful waste. In some cases rights should be defined in terms of short-term rates of discharge, in others in terms of accumulated discharges, so continuous or at least frequent periodic measurement is required. We need liquid waste measuring devices equivalent to the common water meter—on-site, rugged and precise. For most important water pollutants, such equipment was either unavailable or costly in the early 1980s, with apparatus costs per installation ranging from a few hundred to several thousand US dollars (Russell, Harrington and Vaughan, 1986: 76–86).

No one will pay anything for a right which cannot be enforced, and this requires policing of illegal discharges and measurement of the discharges of all right holders, to ensure that individual contributions do not exceed entitlements. Thus effective marketable rights involve accurate measurement of the particular liquid waste at all sewer intake and exit points.

If the measurement costs involved in policing marketable rights are too high, it will not be worthwhile to create them. The introduction of marketable rights is more likely to be cost-effective for large-volume sources of liquid wastes, such as industrial plants, sewage treatment plants, and major commercial developments. While measurement costs remain high, it will be too costly to introduce marketable rights for households and small businesses. This is the pattern we observe in the progressive introduction of charges for liquid wastes by sewerage enterprises in Australia and overseas, and by European environmental agencies which impose liquid effluent charges; volume- and concentration-based charges for industry and sewage treatment plants, and small-customer charges based on flat fees, or water usage, or a combination of the two.

Other problems of implementing marketable rights
There are further technical complications. The waste assimilation

capacity of the environment will vary with seasonal and climatic conditions; to maximise efficiency of waste disposal, discharge rights should vary seasonally, for example. Where individual pollutants interact in the pipe or the environment, setting upper limits to discharges of individual wastes may be less important than ensuring appropriate mixes through time. In some cases the need for control of a sewage constituent may be unclear due to scientific uncertainty. So a once-and-for-all allocation of rights based on imperfect information about the impact of sewage on the natural environment may lead to inefficient use of the environment.

All the above problems apply in any pollution control regime involving quantitative limits on discharges, whether or not the rights so created are formally recognised and tradeable. The only difference with tradeable rights is that if the environment protection agency gets the limits wrong the government may have to purchase rights from owners. Under the present 'command and control' regime it is less likely that government decision makers bear any of the costs of incorrect decisions. This raises the question of whether, in a situation of uncertainty, the agency will act too conservatively in setting discharge limits, which will in turn influence the market values of waste disposal rights.

Markets work best, in the sense of signalling true costs and benefits and transferring property to those who can use it most productively, if there are lots of potential buyers and sellers. If we create marketable rights to sewer disposal of particular liquid wastes, will the rights be traded? Hahn (1989) reports little trading in most water and air pollution permit programs in the US. He attributes this to small numbers of potential traders, to actions of rights holders to restrict competitors, and to difficulties in monitoring environmental impacts, leading to agency restrictions on trading of rights. Thus, when contemplating the creation of marketable rights for a particular liquid waste in a particular sewer system, we must consider the likely number of traders, and the possibility of strategic behaviour towards commercial rivals, for example, one food processing company not selling rights to another to prevent its establishment in the same farming area.

Productive trading may occur between only two interested parties; a rural industry producing organic wastes and a neighbouring town could trade BOD disposal rights so that the job of BOD reduction would be undertaken by whichever could do the job most cheaply.

Conclusions

Market prices for liquid wastes provide information about the costs and benefits of wastes to other people, and the incentive to respond to others' needs. Decisions about liquid waste production and disposal in Australia make little use of prices, and thus the decisions of different individuals along the sewer pipe are not well coordinated, in the sense that individuals are making decisions in ignorance of the benefits and costs of their choices for others. As a result, we do not get the correct mix of liquid waste production, transport, and treatment for permitted discharges into the environment.

Creating markets in sewage wastes will be costly and will take time to implement, but, over time, the costs of failure to communicate costs and benefits along the sewer pipe will almost certainly be much greater. At present, the environmental protection agencies setting sewer discharge limits have no clear idea of the costs that an incremental tightening of standards will impose on industry and treatment plant operators. Individual liquid waste producers and processors have little idea of others' willingness to pay for waste disposal, and costs of providing disposal. Thus, at present, opportunities for desirable adjustments and profitable deals are being missed. And due to a lack of bottom-line rewards for innovations in waste production and treatment, firms are less likely to be using least-cost technologies. Nor is there much domestic reward for Australian firms which design waste treatment and monitoring equipment; their best markets are overseas, where more market-oriented approaches to sewerage have found favour.

Government liquid effluent charges would be a partial solution to the signalling and incentive problems. Corporatisation or privatisation of sewerage, so that sewerage enterprises become net wealth maximising businesses concerned to increase revenues by offering more valued services and to minimise operating costs, would also improve signalling and incentives. These measures would not solve the longer-term problem of uncertainty about politically and bureaucratically determined waste disposal rights and charges. Incentives for long-term research and development and investment in liquid waste reduction, treatment and measurement technologies and installations depend substantially on a clear definition of the future rules of the game in liquid waste disposal.

Marketable liquid waste discharge rights would provide the security and incentive required for greater private innovation and investment in liquid waste disposal. Marketable rights equal to the present permitted levels of discharges accomodate the present quantity standards approach to pollution control. Discharge rights will end up with those who value them most, and thus the overall cost of achieving permitted discharge standards will be minimised.

The high costs of accurate measurement of liquid wastes at all sewer inlet and outlet points are a major impediment to the introduction of marketable rights. As with existing effluent charge systems, implementation of marketable rights would be piecemeal, starting with those sewer systems and types of waste which offer the greatest net payoff, based on unit costs of installed measuring equipment, numbers of waste producing and treatment firms and of potential traders, potential for innovations in waste production and treatment and the state of knowledge about environmental impacts at the sewer outlet. High costs of measurement would rule out the early allocation of marketable rights to households and small commercial businesses. Industrial waste producers, waste treatment firms, sewerage system operators, large commercial establishments and developers who are able to design waste reduction technology or contract out waste treatment, are, however, potential owners and traders of rights, since the quantities involved could justify the installation of expensive measuring equipment.

A measurable industrial waste produced by a substantial number of firms in a large metropolitan area would be a suitable candidate for the initial creation of marketable rights. Providing that all wastes go to the same sewer outlet, and sewer capacity is adequate, trading and the consequent shifts in the location of discharge rights will not alter the environmental impact of waste discharges at the outlet. In the absence of practical experience of marketable rights with sewerage, such institutional innovation is required to demonstrate the advantages and problems involved.

Since marketable rights are proposed on the basis of existing imperfect quantitative discharge standards, environmentalists may complain that the creation of rights to discharge pollutants confirms the existing situation, and contributes nothing to the solution of our pollution problems. This ignores the fact that marketable rights and effluent charges create incentives for market participants to reveal additional information about the costs and benefits of pollution. Providing the measurement problems can be overcome (and this too requires deliberate creation of incentives), the establishment of private rights to sewer disposal of wastes will increase the information base for future environmental decision making, and this has to be good for Australians.

8 A case study — establishing property rights to Chatham Islands' abalone (paua)

Executive summary

The Chatham Islands lie some 800 kilometres off the east coast of mainland New Zealand. Fishing and farming are the lifeblood of the 800 or so inhabitants, with the fishing having experienced some spectacular booms and busts. Seals, whales, cod, and, most notoriously of all, rock lobster, are all fisheries that have boomed and then crashed as the resource was fished out.

A new and lucrative fishery on the Chathams now centres on the shellfish known as paua or New Zealand abalone. The fishery was developed in the early 1970s, and since then has been subject to numerous regulatory attempts to beat the boom and bust cycle. In 1987 paua divers were presented with Individual Transferable Quotas (ITQs) that allow the taking of a specified tonnage of paua each year. The ITQ is in effect a property right in the Chatham Islands' paua harvest. The experience of ITQs in the Chatham Islands' paua fishery thus provides an interesting property rights case study.

The case study is presented in seven sections. The first section makes clear that over-fishing is a consequence not of free enterprise but of fisheries being owned in common by all. The common view that over-fishing is caused by fishermen chasing short term profits is a mistaken one. This mistaken view leads to the conclusion that fisheries, and indeed all resources, should be regulated. However, government regulation is not the answer. As the second section makes plain, the great multitude of regulations that government introduced to conserve paua failed to ensure sound resource management.

The third section of the study describes the introduction of the ITQ system. The ITQ system is a great advance over previous regulations but it is far from ideal. In particular, it does not provide property in the paua stock but only in the paua crop. The fourth section explains problems arising because the property rights are so attenuated. The fifth section describes the political contest that has arisen over paua and which has shifted ashore the wasteful competition that the ITQ system stopped at sea. The sixth section describes the conversion of paua divers through the ITQ system from paua plunderers to paua farmers. The seventh and final section concludes the case study by explaining how the problems that the Chatham Islands' paua fishery confronts can be overcome by allowing a better definition of property rights.

Every-one's property is no-one's care

The blame for over-fishing is usually placed at the door of unregulated free enterprise. Fishermen are seen chasing short term profits, without regard to longer term consequences, with over-fishing the inevitable result. The conclusion follows that fishing must be regulated. However, over-fishing is not a consequence of fishermen chasing the dollar. It is instead a consequence of fisheries not being in private ownership. The problem is the lack of the very institution that lies at the heart of the free enterprise system.

The European tradition is for the sea to be owned by everyone, and the fish in the sea to be owned by no one. 'Roman law laid down that the sea is *communis omnium naturali jure,* or by nature, common to all mankind, and not susceptible of possession in the same manner as land' (Swarztrauber, 1972: 10).

Traditionally the sea has been open to all, with those taking fish, taking ownership of them. 'The fish of the sea ... belong by an absolute title to him who first succeeds in obtaining possession of them' (Fitzgerald, 1966: 433-434)

The reason why sea-fisheries historically were left open to all and not reduced to private ownership was that they appeared limitless. 'The moral reason why ownership is not suitable to the sea is drawn from the consideration that its use is inexhaustible and therefore sufficient for the general service of all, so that it is idle to wish to assign parts of it to individuals' (Pufendorf, 1688:

561).

Unfortunately, fisheries have now proved all too exhaustible, and the result of having them open to all is that they are overfished (Gordon, 1954). The essence of the problem is captured in phrases such as, 'Everyone's property is no one's property'; 'Get it while the getting's good'; 'Why should I save it, if my neighbours will just use it up?' The phrases evoke an image of greed, plunder and careless use, the root cause of which is open entry. When entry is open, resources are overused, used too quickly, and the ability of natural systems to replenish themselves is destroyed.

The reason why is not hard to fathom. Imagine yourself a paua diver. No one owns the paua beds you fish; you simply own the paua that you take. Imagine further that the fishing pressure is such that the paua beds are in danger of being fished out. Your future income from paua diving is thus at risk.

To sustain the fishery you might be willing to limit your catch and forgo some of this year's income to preserve next year's. However, although you can restrict your catch, you cannot ensure that the other divers do likewise. They can keep on fishing, and, indeed, they can take the very paua that you leave behind. The other divers can profit at your expense and reduce your efforts at conservation to futile self-sacrifice. There is nothing you can do to conserve the resource. Instead of leaving paua behind you may as well fish as hard as you can while you can. The same goes for each and every diver. The logic of open entry drives you and the other divers to fish the paua out.

Now consider your behaviour if through some quirk of biology paua lived ashore on your own land, or alternatively, if through some legal oddity paua beds were privately owned and you owned a slice. You would not fish out your own paua; to do so would simply be to strip your own private asset. Instead of fishing your paua out, you would husband them. If you owned a paua bed, every incentive would exist for you to maintain the paua; to do otherwise would be to jeopardise your future livelihood.

The essential feature of private ownership is exclusive use (Cheung, 1970). If you own the paua bed, you have an exclusive right to the paua. You can prevent other divers from taking your paua and thereby control the harvest and manage the resource.

If entry is open to all, your right to take paua is non-exclusive. The right of exclusive use is also what makes the paua bed your asset. Anyone who wants to take your paua has to buy or otherwise negotiate the right to do so. The resource thus attracts a price, and it is your private asset. To maintain the value of your asset, you must not over-fish. Private ownership thus gives you the authority to manage the resource while making you responsible for your management.

Resource conservation through government regulation

The fisheries experience in New Zealand has certainly not been one of private property and free enterprise. The experience instead has been one of common property and state regulation. Common ownership has been maintained and fishing regulated in an attempt to counter over-fishing.

The paua fishery was for a long time open to all and unregulated. The first regulation applying to the taking of paua came as part of general fishery regulations. The Fisheries Amendment Act of 1945 to the Fisheries Act of 1908 required fishermen fishing commercially to be licensed. The move was in part a conservation measure. The Sea-fisheries Licensing Authority in considering any application for a licence was required to have regard to 'the desirability in the public interest of conserving sea fisheries.' Thus, since 1945 anyone taking paua for sale has been required to have a licence.

However, back in the 1940s there was little commercial interest in paua. The resource was conserved simply because demand was low. As O.M. Moore (1982: 33), a senior fisheries officer with the Ministry of Agriculture and Fisheries, recalls:

> Before the late 1960s the taking of paua, both by amateurs and commercial fishermen, caused few problems. Stocks were generally high, especially in the less accessible areas of Southland (including Stewart Island), Marlborough, and Wairarapa, and they were virtually untouched in the Chatham Islands. The amateur take was low, as paua was then considered an acquired taste, except among the Maori people, for whom it is a traditional food. Commercial ex-

ploitation was confined to a few small-time operators, as the only export market available at that time was Australia. There raw, black paua meat was further processed and bleached. The raw New Zealand product fetched only low prices and as a result attracted few commercial operators.

The pressure came on the paua resource in the late 1960s. In 1968 an entrepreneur discovered how to bleach the black paua meat and thereby penetrate overseas markets. Moore (1982: 33) recalls the resulting price hike:

> In 1968 the situation changed suddenly when the Palmerston North based company of Prepared Foods Ltd. discovered a satisfactory bleaching process for New Zealand paua and was able to market and export the fully processed and canned product under the name of New Zealand abalone. The price for raw paua increased by about 50%.

The race for paua had begun. 'Many persons saw the opportunity of making big money and entered the fishery' (Moore, 1982: 33). The Wellington area took the initial brunt. As catches started to decline divers moved further afield to Wairarapa, Marlborough, the west coast of the South Island, and eventually the Chatham Islands. In 1972 over 400 diving permits were issued. Paua stocks in many areas were seriously depleted.

In an attempt to conserve the fishery the government set size limits and gear restrictions. A size limit of 5 inches (125 mm) was introduced in 1971, and the use of underwater breathing gear banned. Areas of the coast were also closed to commercial divers. Catch quotas for amateurs were also introduced and progressively tightened. The amount one person was allowed to take in a day was restricted to a four gallon measure in 1959; reduced to two gallons in 1968; and to 10 paua in 1972. In heavily fished areas daily and weekly catch quotas were introduced for commercial divers (Report of the Fishing Industry Committee, 1972: 35). However, the difficulty of policing the quotas rendered the restrictions useless and the entire paua fishery was closed for four months in 1972.

The fishery was opened once more to year-round fishing the following year. In that year a quota on the export of paua meat was introduced as a conservation measure. 'The export quota for meat was divided proportionately among established export companies only. The amount of quota depended on the quality and quantity of the company's product and the time it had spent on marketing paua' (Report of the Paua Shell Review Committee, 1988: 11). Five companies received quota. The product also was regulated. 'Exports were restricted to bleached, canned meat as the raw, black product was unacceptable to some overseas buyers for religious and cultural reasons' (Report of the Paua Shell Review Committee, 1988: 11) Overseas buyers interested in paua other than bleached and in a can could not be supplied. Between 1973 and 1984 the export quota for all New Zealand was held at 71,400 dozen cans of processed product, or about 928 tonnes (whole landed weight) of paua (Murray and Akroyd, 1984). No quota applied to domestic sales.

A hotch-potch of regulations was thus applied to the taking of paua. Regulations applied to who could fish (commercial divers had to have a licence), how you could fish (underwater breathing gear was banned), what fish you could take (paua had to be above legal size), how many fish you could take (catch quotas were in force), and where you could fish (areas of the coast were closed to commercial divers). Regulations also applied to who could export paua, how you could export it, and how much you could export.

The regulations restricted the taking of paua but in so doing they restricted the potential of the fishery. Licence requirements locked out new entrants, gear restrictions kept the cost of taking paua high, and export quotas discouraged product development and price competition. The regulations were themselves a cause of considerable waste.

What is more, the regulations never got to the heart of the problem. The fishermen were left fishing a common stock and total catch was never directly constrained. The outlook for the paua fishery remained one of ever-increasing effort for ever-fewer fish. Any rent generated was always to be dissipated by divers competing amongst themselves for a share of the resource.

The hotch-potch of regulations did not and could not provide the necessary incentives to ensure sound resource management.

Establishing property in paua

The hotch-potch of paua regulations was swept away when the Individual Transferable Quota (ITQ) system was extended from the inshore fisheries to paua. Within the ITQ system the total catch is constrained by setting a Total Allowable Catch (TAC) that limits the amount of fish that may be taken in a year. To prevent fishermen wastefully 'racing' against each other to get as large a share of the TAC as they can, the TAC is divided up into individual catch quotas. The individual catch quotas serve to allocate the TAC amongst individual fishermen and fishing firms before the fishing season gets underway.

In New Zealand the quota were allocated in perpetuity, the intention being to give fishermen long term security. In order not to lock quota holders into the fishery, and to lock non-quota holders out, the quota were made tradeable. Trade in quota allows fishermen and fishing firms to enter and exit the fishery freely. The ITQ system thus attempts to control the total catch while leaving fishermen and fishing firms free to catch their quota as they best see fit.

The difficult step in establishing the ITQ system is the initial allocation of quota. In the paua fishery ITQs were allocated on the basis of catch histories. Divers who were licensed, and who worked the fishery, received quota. Beginning in 1982 there had been a moratorium on the granting of licences to take shellfish, including paua. In 1985 Individual Quota (IQ) (not tradeable) set at 90% of a licence holder's catch in either 1982 or 1983 were allocated. Initially 227 tonnes went to 35 divers in the Chathams. A further allocation of 27 tonnes was made by a Ministry of Agriculture and Fisheries committee for licence holders who did not have a catch history for 1982 or 1983, or for whom those years were not a reasonable reflection of their catch history. As a result, a new total of 254 tonnes was established.

In June 1985 a Paua Quota Review Committee was established to report on objections received from paua divers. The Committee recommended further increases in IQ totalling 29 tonnes and, in addition, the Ministry of Agriculture and Fisheries itself allocated a further 49 tonnes to take the total to 332 tonnes.

Legislative backing for the ITQ system was enacted as an amendment to the Fisheries Act of 1983 in 1986, with full implementation of the ITQ system for paua scheduled for the 1st October 1987. Another 35 tonnes were allocated in the move from IQs to ITQs to give a new total of 367 tonnes. All quota holders then took a 28.6% quota cut to meet the preset TAC of 262 tonnes. Taxpayers funded this later cut to the tune of $5,400 for each tonne divers gave up with taxpayers contributing $1.4 million nationwide to buy back 260 tonnes of paua quota.

The allocation of ITQs was further complicated when Maori interests obtained a High Court injunction preventing the establishment of ITQs for Maoris fishing paua, and other species. Maori representatives contested the authority of the government to allocate individual property rights in fisheries when ownership of the resource had not been established. The High Court ordered ITQ to be issued subject to the condition that the quota terminate if at any stage Maori tribes proved in court that the fish belonged to them, or if the Crown acknowledged that Maori tribes had control of the fishery. All but eight tonnes of the original ITQ issued for paua in the Chathams are subject to the High Court order.

The initial allocation of ITQs in the Chatham Islands' paua fishery provided 37 divers with a total of 261 tonnes of quota (yes, one tonne was lost in the system!). However, as part of the implementation procedure for the ITQ system, the Fisheries Amendment Act 1986 established a Quota Appeal Authority (QAA) to which any person could appeal. As of February 1990 the QAA had allocated a further 21 tonnes of paua quota to the Chatham Islands giving eight divers additional quota and two divers an initial allocation. The TAC for Chatham Islands' paua has thereby been increased to 282 tonnes.

Following the introduction of the ITQ system the earlier paua regulations were rationalised. The only restrictions that remain are the size limits and the ban on the use of underwater breathing gear. The market for paua is now competitive and product development is free of the regulations previously intended to protect the resource. The freeing up of the market has had a considerable effect. In 1984 divers were getting only $1 a kilogram greenweight; by December 1989 they were getting $15.

These high paua prices spilled back into higher quota prices when in December, 1987 permission was given for trade in paua quota. This trade has been heavy. After all, existing licence holders had effectively been locked into the fishery since 1982 and new divers locked out. One hundred and eighty tonnes of Chatham Islands' paua quota changed hands between February 1988 and March 1990. Ten divers sold out entirely, and 17 new divers entered the fishery. Paua quota now sell for around $40,000 a tonne, and are very valuable assets.

ITQs as attenuated property rights

Although the ITQ system is a vast improvement over earlier regulations, it is far from ideal. Two problems, in particular, stand out in the paua fishery. The first is that the TAC must sometimes be reduced to protect paua stocks, or, perhaps increased to take advantage of stock improvements. Adjusting the TAC is both difficult and contentious. It is difficult because the government officials making the adjustments do not know the resource as well as the divers doing the harvesting, and it is contentious because the costs fall upon the quota holders and not the government officials doing the adjusting.

The original plan had been for government to adjust TACs by buying and selling quota by tender (Ministry of Agriculture and Fisheries, 1984: 11). In the event the Fisheries Amendment Act of 1986 that set up the ITQ system provided for the proportionate reduction of all ITQ, with the Government compensating quota holders for the 'fair market value of the individual transferable quota.'

The mechanism for TAC adjustment was changed yet again with the Fisheries Amendment Act of 1990 that enables proportionate cuts without government compensation. These changes have shifted the costs of TAC reductions from government to quota holders. They have also considerably undermined quota which are now effectively specified not as a fixed tonnage but as a percentage of an adjustable TAC.

The second problem with the ITQ system is that the rights in question are rights to a part of the harvest from the resource as opposed to the resource itself. The individual quota holder is left without a property interest in the fishery. The incentive therefore exists for individual quota holders to take in excess of their quota and for non-quota holders to poach paua. Under current arrangements the only way to counter 'quota busting' and poaching is to monitor paua divers and enforce quota regulations.

The government monitors each diver's catch by having them fill out on landing their catch a Catch Effort Landing Log which must then be available on demand to any fisheries officer or examiner. Divers must also complete and regularly submit to government a Quota Management Report detailing by area the quantity of fish caught for each species for which quota is owned or leased.

Also, fish processors have to submit regularly to government a Licensed Fish Receivers Return (LFRR). The LFRR shows the quota holder's name, the fisherman's identification number, species, and greenweights for all fish received. It allows the cross-checking of fishermen's logs and reports. The bureaucratic detail needed to monitor each diver's catch makes the ITQ system expensive to run.

Enforcement has also proved a problem. Enforcement officers estimate that six hundred tonnes of paua are taken illegally in New Zealand waters each year. The Total Allowable Catch for all of New Zealand is only 1114 tonnes and illegal fishing is seriously undermining the ITQ system.

The problem of enforcing quota is particularly acute in the Chathams. 'In the Chathams the surveillance is currently provided through one resident officer who has the task of enforcement for the seas around the many Islands of the Chathams group' (Taylor Baines and Associates and Lincoln International, and Associates and Lincoln International, 1989: 116). The resident officer is provided with a 4-wheel drive vehicle and a pair of binoculars, yet he must police not just paua but the entire Chatham Island's fishery.

Fishing political waters

Illegal fishing is not the only threat the fishery confronts. The high returns and high quota prices have caught the attention of envious eyes and various groups without quota are lobbying government for a slice of the pie. Indeed, government is one of the groups after a slice. In establishing the ITQ system for managing fisheries around New Zealand, the government set the precedent for extracting the rent the fishery earns. This was done by establishing a nominal rental or tax on quota. 'Government policy has been to gradually increase resource rentals until the value of annual traded quota approaches zero' (Clark & Duncan 1986: 118). The stated purpose of increasing the tax on quota is to encourage new fishermen, discourage speculation, and return the rent the fishery earns to the people of New Zealand (Clark, Major & Mollet 1988: 334). The tax at present is only $131.54 a tonne on paua quota, and it would have to be increased substantially if quota prices were to be driven from $40,000 a tonne to zero.

Chatham Islanders without fish quota are also after a slice of the paua pie. A recent study of the Chatham Islands economy recorded a number of submissions that argued 'that due to the location of the Islands, any economic rent derived from the fishing resources with 'the Islands' 200 mile EEZ (Exclusive Economic Zone), should be returned to the Islands' (Taylor Baines and Associates and Lincoln International, 1989: 121). The idea again is to tax away the value of the quota but in this instance to direct the revenue to a political body on the Chathams. The submissions noted that the revenue could be used for local administration funding, fishing resource management, and Island-based investment.

'Another common submission was for the Island to either own or at least hold control over the fishing resource, so that the allocation of the resource quotas could be directed by the Islanders only to residents' (Taylor Baines and Associates and Lincoln International, 1989: 121). The idea is for quota rather than fishery revenues to be taken and then reallocated to locals. A related proposal is to require quota holders to live in the Chathams. 'Yet another aspect of those submissions was the proposal to apply residential clauses in order to get people to invest on the Island and to have the fish landed on the Island to create increased economic activity' (Taylor Baines and Associates and Lincoln International, 1989: 121).

In addition, three indigenous groups are using the Treaty of Waitangi to claim the paua resource. The Treaty of Waitangi, signed in 1840 by representatives of Queen Victoria and Maori tribes, established British governance over New Zealand but also affirmed the rights of the tangata whenua (original people) to their lands, forests, and fisheries. The Treaty of Waitangi Act of 1975 established a Waitangi Tribunal to investigate Maori claims against the Crown, while the Treaty of Waitangi Amendment Act of 1985 has enabled the Tribunal to examine claims dating back to the signing of the Treaty.

The first indigenous claim arises from the arrival of the Taranaki Maoris in the Chathams in 1835 and their subsequent conquest and subjugation of the original inhabitants of the Chathams, the Moriori. The claim attests that this conquest puts the Chatham Islands within Taranaki tribal boundaries. The Taranaki Maoris claim exclusive title to possession, use and control of the fishery. Where such rights cannot now be restored the claim is for sufficient compensation for loss of mana and of such rights. The Taranaki claims are made on behalf of the beneficiaries of the Taranaki Maori Trust Board and associated corporations and trusts.

The second indigenous claim is from the Moriori, and is not so straightforward. The Moriori, spurred to action by the Taranaki claim, assert that as the indigenous tribe of Rekohu (Chathams), they have standing as tangata whenua before the Waitangi Tribunal and are therefore justified in presenting a case for the return of the fishery resource to the Chathams. The original intent behind this proposal was that the formal structure under which the Chathams' fisheries would be owned and managed would be along the lines of an Iwi (tribal) Authority. But a later proposal was that an independent body be incorporated to govern the management and control of the fisheries, with representatives of Maori, European and Moriori participating in management. Supporting the claim of the Moriori to rights as indigenous people is the Moriori Tchakat Henu Association of Rekohu. The Association has amongst its objects the promotion of the wise management, conservation and administration of natural resources in Rekohu, and it consists of members committed to the objects of the Society. These members are either resident members (whose principal place of residence is Rekohu), non-resident members (able to take up voting rights if their principal place of residence is Rekohu for more than three months in any one year), or special members (under 18 years of age or yet to complete the residency requirement).

The third indigenous claim is from Te Rununga O Wharekauri Rekohu, the Chatham Islands' Maori, who assert that they represent the tangata whenua of the Chatham Islands by descent and who accordingly claim exclusive title to possession, use and control of their fisheries.

In response to the many claims upon the resource a recent government-sponsored *Review of the Chatham Islands Economy* argued for 'both Government officials and fishers to compromise their extreme positions and overcome their resistance to sharing power and responsibility' (Taylor Baines and Associates and Lincoln International, 1989: 118). The consultants proposed the setting up of a Chatham Island Fishing Authority to represent the various interests. 'The representation on the Fishing Authority would come from the following groups: the fishers through their respective speciality groups; the processors; the service sector, including MAF fish, Fishing Industry Board (FIB), private consultants etc; the CICC (Chatham Islands County Council) chief executive officer, the CDC (Community Development Corporation) chief executive; and the Iwi Authority' (Taylor Baines and Associates and Lincoln International, 1989: 119). The Authority's task would be to manage the fishery with the rent from the fishery being split between central government and the Authority. The proposal is to get the many interests represented in management and to manage the resource and direct investment politically.

Notwithstanding the ITQ system, the paua fishery remains up for grabs. Both central and local government are after a slice, as are local Chatham Islanders, and the three indigenous groups.

There are now more people than ever fishing for paua. The fishing effort is not now going into trying to take paua ahead of another but rather to out-lobby other groups after the resource. The wasteful competition that the ITQ system stopped at sea has simply been shifted ashore. In the process neighbour is being set against neighbour. One group's gain will be another group's loss.

From hunter to farmer. The various attempts to re-slice the pie politically are not only proving wasteful and divisive, but are also undermining the ITQ system and scaring off the very investment that could make the pie even larger. The investment made back into the resource follows from a feature of the ITQ system that is often overlooked. Quota holders are in effect a club with the exclusive right to take paua commercially. Club members hold a valuable property right in the paua harvest in the form of their ITQ. It is in the club's interest to maintain and, if possible, enhance paua stocks, in order to maintain and improve the income stream generated from the fishery and hence quota prices. Accordingly, the Chatham Islands' paua divers have formed the Chatham Islands' Shellfish Reseeding Association which has amongst its objectives, 'to foster and promote the enhancement of the fishery stock in the Chatham Islands,' and 'to foster and promote scientific research into shellfish programmes in the Chatham Islands.'

In an attempt to enhance the fishery the Association has contracted the Ministry of Agriculture and Fisheries to conduct a trial paua seeding programme in the Chathams for the year ending 1st October 1990. The contract involves rearing 100,000 paua to a size of about 7 mm, seeding and monitoring these juvenile paua, and reporting on the success or otherwise of the trial. The trial is funded through a 1% levy upon Association members' paua sales for the year. Thirty-four out of the 46 quota holders on the Chathams have joined the Association and these 34 hold 84% of the Chathams' TAC. If the trial proves successful, the paua divers are set to seed the paua fishery around the Chathams on a large scale.

The exclusive right to harvest the resource as secured by the ITQ system has thus converted Chatham Islands' paua divers from resource plunderers to custodians of the resource. Unfortunately, the insecurity generated by present ITQ management and the many competing political claims are undermining investment and conservation. Roger Beattie, the secretary of the Reseeding Association points out the problem. 'Paua divers are not going to invest in conserving the fishery unless their rights to harvest where they have sown are secure. Farmers would not put fertiliser on their land unless they were sure they could reap the benefit. Paua divers are no different to farmers. They will only invest back into the resource if their property rights are secure.'

Closing the commons

The problems confronting the Chathams' paua fishery are plain enough. The rent captured through the ITQ system is in danger of being dissipated in a political contest over the resource. At the same time, the investment that could enhance the resource to everyone's advantage is being scared off. Although paua divers are investing in the paua fishery, they are investing less than they otherwise would be. On top of these difficulties, the TAC and hence quota are able to be adjusted by government officials who have no financial interest in the fishery, and the ITQ system is being further undermined by the bureaucratic detail needed to monitor each quota holder's catch and by the government's inability to enforce quota regulations.

The claims to the resource from non-quota holding Chatham Islanders are counterproductive. They have as their stated aim the promotion of local investment. However, the way to promote investment in the Chathams is to secure the quota. Moreover, the various claims all involve replacing private investment with political investment. Political investors do not risk their own capital and are motivated to seek out the most politically attractive investments, not the most profitable. The substitution of political investment for private investment will harm rather than benefit the Chathams' economy.

The claims lodged with the Waitangi Tribunal are not going to go away, but it would certainly help to secure the ITQ system if the government acknowledged that these claims are against the Crown and not existing quota holders. The government should state categorically that if claims to the Waitangi Tribunal are recognised, government will not ride roughshod over existing quota holders' rights. That is to say, the government will neither take quota from existing quota holders nor increase the TAC to allow more quota to be handed out. If government does agree to give quota to claimants, existing quota holders should be bought out at the market price. This would serve to secure quota and perhaps minimise resentment by dispersing the costs of reallocation.

Resource taxes are also a threat. Quota cannot be secure so long as government plans to expropriate the rent by driving quota prices to zero. The quota can be made secure by government indicating that it has no intention of raising the quota tax and the best way of doing this is to abolish the tax completely. The quota tax serves no good purpose, being a very small source of revenue, and any move to increase it would only serve to undermine the incentive the ITQ system provides for quota holders to conserve and enhance the fishery.

Another contentious issue is the adjusting of the TAC. It would be much better if the TAC were set by quota holders each year instead of government. Quota holders have the detailed information of the fishery that government lacks, and with quota giving them a stake in the fishery, they have every incentive to adjust the TAC if the fishery is being either over-fished or under-fished. The decision would then rest with those who bear the cost. In taking responsibility for setting the TAC, quota holders would no doubt seek out and employ fishery scientists who could assist them in making the decision. Fishery scientists would then work advising quota holders who have a direct financial stake in the fishery.

The setting of the TAC by quota holders would require some form of voting procedure, and it would be preferable if voting rights reflected the extent of the quota holder's stake in the fishery. That is to say, voting rights would be set according to the amount of quota held. If each quota holder had but one vote the influence of each person's vote would be disproportionate to the costs and benefits incurred by that vote.

Monitoring and enforcement also need not be entirely a

government responsibility. Indeed, there is considerable scope for self-policing. There are currently 46 quota holders out on the Chathams spending considerable time lifting their quota and they should be encouraged to police the resource themselves. Although it can profit individuals to go over their quota and steal paua, it costs them if other quota holders (and non-quota holders) do likewise. If everyone exceeds their quota the fishery will be over-fished, paua divers' incomes will fall, and the price of quota will likewise fall. If quota busting and poaching become commonplace, paua divers will find themselves back to the situation of the hunter plundering an unowned resource. Each quota holder will realise that over-fishing is the road to ruin, but that there is nothing he can do about it, and that he may as well be in while he can. It is in the interests of the quota holders as a group to see that the quota are enforced. As in the case of setting the TAC, monitoring and enforcement should increasingly be the responsibility of quota holders. If they don't take responsibility, it is hard to see who will.

Monitoring and enforcement is difficult because it is almost impossible for one diver to know if another is within his quota. One consequence of quota holders taking increased responsibility might be that they agree to carve up the coast to lower monitoring and enforcement costs. This would involve each quota holder agreeing to take his quota from a specified area. Anyone taking paua from outside their area would be taking them illegally. Of course, the productivity of the resource in his area would have to be in proportion to quota held. If a quota holder has five tonnes, he should be given an area equivalent to his five tonne holding. The carve up could presumably proceed privately and voluntarily.

The incentive to carve up the coast follows from quota holders taking increased responsibility for monitoring and enforcement. However, the advantage of such a carve up does not end with reduced monitoring and enforcement costs. Specifying rights to an area of coast would allow quota holders to decide as individuals, or as groups, their commitment to investing in enhancing the resource in their area. It would also allow those who want to invest greater sums in enhancement to do so, and to reap the reward. The increased reward would come from their being able to more easily fill their quota, and, perhaps in the not too distant future, to increase the harvest from their area. At present, investment in enhancement is tied to what the most unwilling member of the Association is prepared to commit.

Indeed, if the coast were carved up the ITQ system for paua would become redundant and the paua beds would be effectively managed by private owners. The owners would have authority to prevent over-fishing in their area, and, so long as they remained free to sell the rights to their paua, they would have every incentive to conserve their resource rather than plunder it. A TAC would not be needed. Private owners would simply decide their own level of harvest in their own area, just as farmers decide the stocking rates on their own land. Monitoring and enforcement also would not prove such a problem. No government monitoring of the harvest would be necessary, and enforcement of exclusive rights to the paua in a given area would prove no more difficult than the enforcement of exclusive rights to stock on land.

Conclusion

The Chatham Islands' paua fishery illustrates the importance of establishing property rights in resources to ensure sound resource management. The attempt to regulate fishing in the absence of property rights was a failure. The introduction of the ITQ system not only served to constrain the total catch directly, but also provided the incentives necessary to convert paua divers from resource plunderers to resource custodians. The fishery also illustrates the dangers of the property rights approach. The potential gains in establishing property rights can easily be dissipated in the political arena as interests lobby for a slice of the pie (Anderson & Hill, 1983). There is no easy answer to the problem. But clear, sharp and rigidly applied rules to guide a quick and final allocation are certainly important. One mistake made with paua ITQ was to stretch the rules to the limit and to continue allocating quota to divers after the initial allocation. Another mistake made was to encourage the many claimants to lobby for the resource and to re-establish the common property problem within the political arena.

The Chatham Islands' paua fishery also illustrates that the ITQ system should not be viewed as an end point of reform. As Anthony Scott (1988: 289) has explained:

> An ITQ harvesting regime, requiring continued regulation, is best seen as only a brief stage in the development of management. Its evolution can be expected to continue until each owner has a share in the management decisions regarding the catch, and, further still, until he has an owner's share in management of the biomass and its environment.

The incentives and opportunities the ITQ system provides can thus be built upon by allowing quota holders to club together for the purpose of fisheries management. As a club, quota holders can control their fishery and can capture the benefits of conservation and investment. In establishing an exclusive right to the fishery, albeit a collective one, the problem of over-fishing has been solved.

The evolution of the ITQ system is well underway in the Chathams' paua fishery with quota holders seeding the fishery themselves. The prospects for a profitable and sustainable fishery appear bright. The danger is that present policy is directing the energies of the Chatham Islanders away from conservation and investment and into unproductive competition over the resource.

9 Air pollution

Executive summary

Air pollution is an inevitable consequence of living. It can be eradicated only at costs which would be totally unacceptable. Air pollution comprises many different elements. As a phenomenum, it is associated with large population concentrations and industrial emissions. It has, however, become much diminished as a cause both of harm and irritations to those potentially suffering its consequences most severely—people located within conurbations. This is notwithstanding continued population growth.

More benign levels of air pollution over recent years can be traced to a number of features:

- *industrial change, whereby a lower proportion of needs in progressively richer societies come from the outputs of 'smokestack' industries;*

- *the dispersion of 'smokestack' industrial facilities away from central city areas to the peripheries of conurbations;*

- *regulatory controls on emissions from individuals (cars and household heat generation) and industrial facilities.*

Countering air pollution presents an unusual challenge to those designing taxes, charges and remedial expenditures. Those who are affected by pollution cannot easily come together so that they may bargain with polluters over a mutually agreeable level of emissions. Moreover, the great bulk of pollution is emitted by numerous sources—and further, the 'polluter' and its individual 'victims' are by and large the same.

Markets, incorporating vested ownership and tradeable rights, are generally the most efficient means of bringing about optimal abatement expenditures. But, given the major problems of monitoring vast numbers of minor sources of air pollution, with present technology, there are at present some apparently insuperable problems involved in developing and monitoring the contractual approaches which can allow market solutions to operate. 'Command and control' solutions remain the best option for these sources.

In the case of major sources of air pollution, the prospects of enlisting market based mechanisms offer much more promise. In particular, tradeable rights in pollution have been demonstrated to save considerable costs where some sources can achieve abatement levels more cheaply than others. For this reason, tradeable rights offer greater flexibility and cost savings than the alternative market based instruments, taxes or charges.

Introduction

Air pollution agents are manifold. Those specifically targeted for control normally include particulate matter (smoke), ozone, sulphur dioxide, carbon and lead.

Over the years the problem of air pollution in western countries has been successfully addressed. In London in 1952, some 4000 deaths resulted from an extended period of air pollution. Today the air is much cleaner, notwithstanding much increased traffic and higher energy generation. Diseases associated with pollution, like influenza, pneumonia and tuberculosis, were responsible for about one quarter of deaths at the turn of the century and now account for less than 5% in a population where life expectancy has increased by over a half. It might be said that the market for death is declining and the market share of pollution related causes falling! To be sure, much of the improvement stems from factors like improved medical treatment, but to a major extent it is due to a cleaner urban environment.

Curiously, one of the patron saints of environmentalism, Paul Ehrlich, also takes the view that air pollution is a readily resolvable problem. Ehrlich's view might be conditioned by his ideological battles within the environmentalist movement in the course of which he has sought to propel population growth to an ascendancy which others have rejected. In his interview in *The Ecologist* (1973) he said:

...from the point of view of an ecologist... (air pollution

is) one of the relatively trivial problems. It is amenable to rather rapid technological cure and is just a symptom of some of the things we're doing, rather than something ecologically serious.

Of course, it could be argued that Ehrlich, who was forecasting widespread famine by the early 1980s as a result of Malthusian analyses of population growth, has been discredited (apart from within a particularly bizarre wing of the environmental movement).

Air pollution in Australia

Air pollution levels in major Australian cities have generally shown an improvement over recent years. In Melbourne, sulphur dioxide levels have trended downwards and in 1988 were less than one third of the maximum acceptable peak levels. Chart 9.1 illustrates peak one hour and 24 hour sulphur dioxide (SO_2) and airborne particle trends for the industrial suburb of Footscray. SO_2 levels in Australia are low by world standards because of the low sulphur fuel used.

Carbon monoxide (CO) levels have also trended down to magnitudes well within the maximum acceptable, although nitrogen dioxide (NO_2) levels have remained relatively close to their 'maximum acceptable level' and ozone (O_3) levels are above those defined as acceptable.

Broadly comparable findings to these were reported for Brisbane by Verrall and Simpson (1988). At the time of Queensland's Clean Air Act of 1963 and the coming into force of regulations giving effect to it, (in 1968), major pollutive industries in Brisbane included four coal fired electricity generating stations, brickworks and a host of other coal and wood burning facilities. Since then, although there has been a six-fold increase in the number of industrial premises, other factors have acted to diminish pollution levels including:

- coal powered electricity generation has ceased within the metropolitan area;
- railways have been electrified;
- domestic burning has been banned.

Chart 9.1
Peak one hour and 24 hour sulphur dioxide (SO₂) and airborne particle trends — Melbourne (Footscray)

SO$_2$ (parts per hundred million) particles (parts per million)

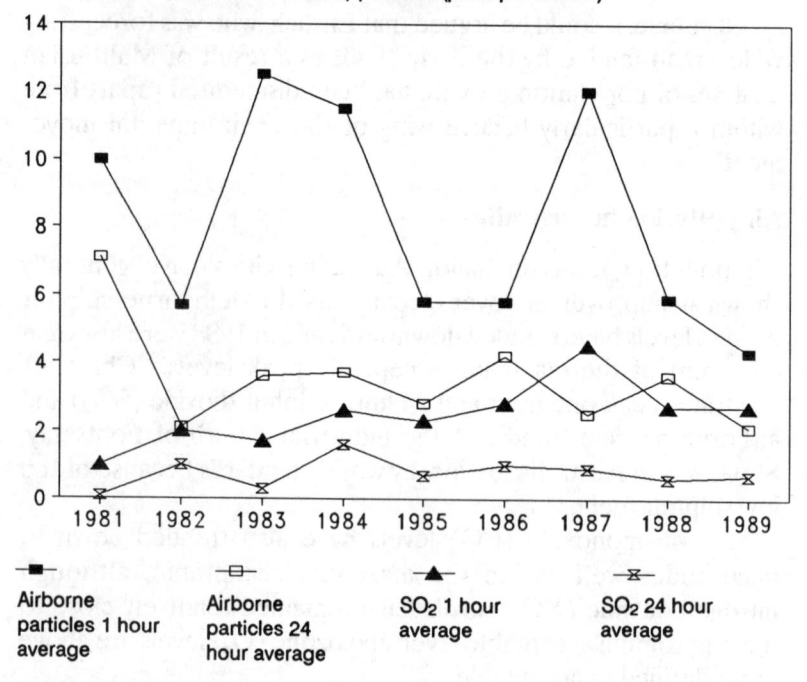

■ Airborne particles 1 hour average
□ Airborne particles 24 hour average
▲ SO$_2$ 1 hour average
✕ SO$_2$ 24 hour average

Source: Vic. EPA

From the late 1970s, ozone levels have remained similar to those observed earlier; NO$_2$, lead and SO$_2$ show slight declines; smoke has declined markedly (and visibility has improved); and CO has shown a major decline (see Charts 9.2 to 9.4).

The abatement of urban air pollution levels have been achieved by 'command and control' regulation. Where markets do not automatically equilibrate supply and demand because of monitoring difficulties, total permitted supply must be specified by a government authority. Such quasi-market approaches will pay dividends when applied to some sources; however, continuation of more directive command-and-control approaches seems to be inevitable in the case of domestic and, perhaps, automotive emissions. In a strict sense, therefore, the achievement of

Air pollution   211

Chart 9.2
Peak eight hour and one hour averages for Carbon Monoxide (CO) levels — Melbourne region

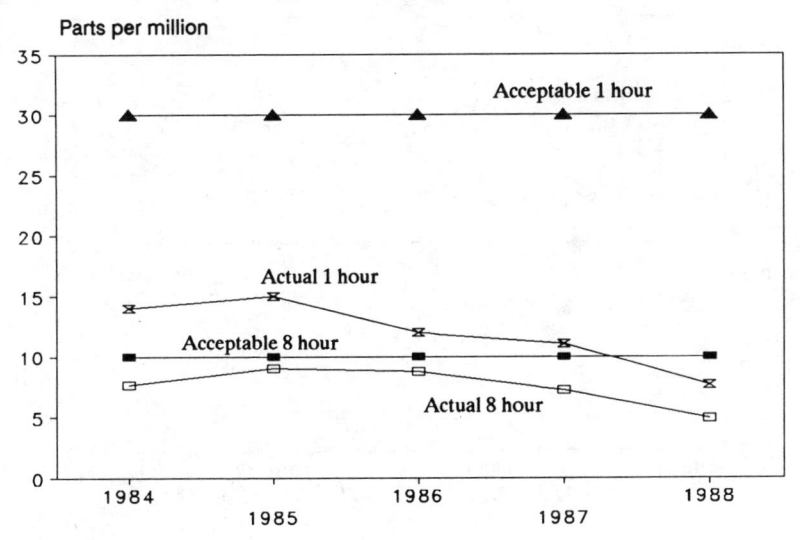

Source: Vic. EPA

Chart 9.3
Peak 24 hour average nitrogen dioxide (NO_2) levels — Melbourne region

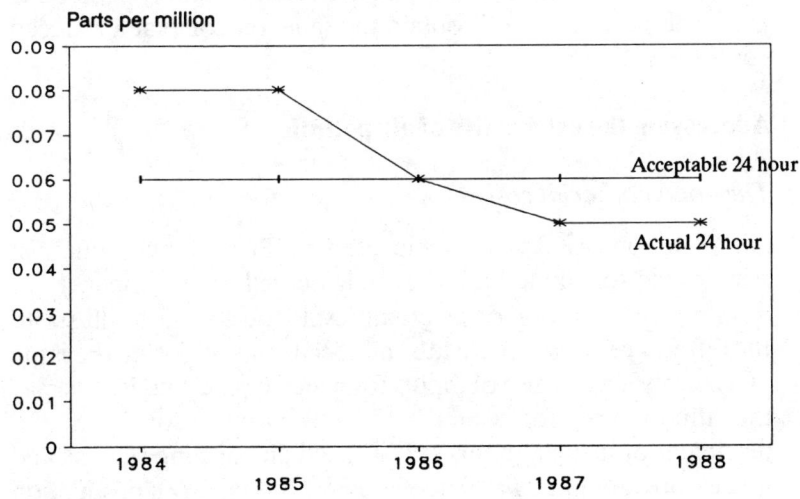

Source: Vic. EPA

Chart 9.4
Peak one hour average ozone (O$_3$) levels – Melbourne region

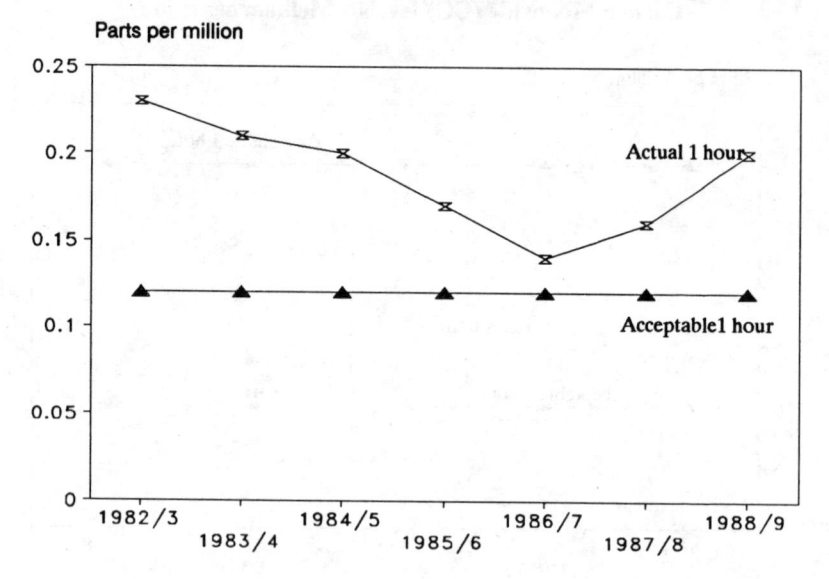

Source: Vic. EPA

efficiency largely turns on the *nature* of the regulation. If market mechanisms are employed to allow polluters flexibility in meeting the levels desired, we can obtain the same outcome at a reduced cost.

Addressing the externality of air pollution

The notion of social costs

Air pollution was the example used by Pigou (the economist responsible for pioneering the study of welfare economics). In developing the notion of externalities, Pigou sought to illustrate the difference between private and social costs by posing the issue of a factory making use of inputs for which it paid, and inputs (say the atmosphere) for which it did not pay but soiled (to the detriment of its neighbours). If it faced diminishing returns and if each private input was valued equally the factory's production

Air pollution   213

Chart 9.5

(a) Positive externalities

With the market demand curve DD, price and quantity are set at P1 and Q1 respectively. However at Q1, the value curve VV which incorporates the positive externalities, yields a price P2 which would be offered. Q2 is the quantity which would be supplied if suppliers could obtain remuneration equivalent to what consumers would be prepared to offer.

(b) Negative externalities

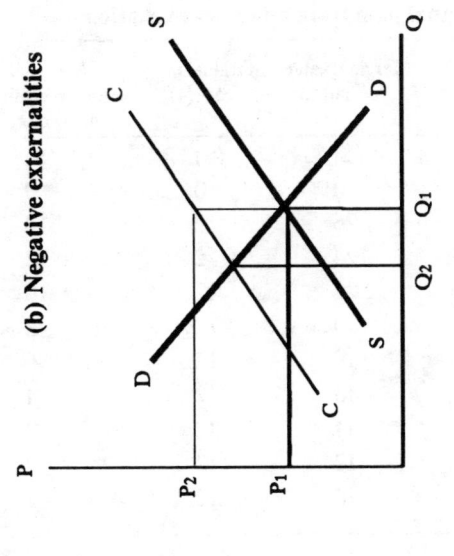

Where social costs exceed private costs, curve SS is replaced by CC. At the private equilibrium quantitiy Q1, the price which offers appropriate compensation at all factors of production is P2. Q2 is the quantity which would constitute equilibrium if suppliers could be appropriately remunerated.

Table 9.1
Private marginal gain from a factory operation

Output (1)	Value of output (2)	Marginal value of output (3)	Marginal input cost (4)	Marginal economic gain to the owner (5)
	($)	($)	($)	($)
0	0	0	0	0
1	26	26	12	14
2	50	24	12	12
3	72	22	12	10
4	92	20	12	8
5	110	18	12	6
6	126	16	12	4
7	140	14	12	2
8	152	12	12	0
9	162	10	12	−2

could be represented by Table 9.1

Under these circumstances, the owner would produce up to the level (8 units) where his marginal costs equalled the marginal value of his inputs.

If uncontracted costs which cannot be charged for are added to this example, and these costs rise at a constant rate with each additional unit of output, then the marginal economic gain to society is different from that of the owner (shown in column 5, Table 9.2).

From his analysis, Pigou concluded that a tax equal to the uncontracted costs should be imposed—a tax which would bring production back to four units.

Freeman, Haveman and Kneese (1973) in Chart 9.5, offer a diagrammatic version of the externalities which Pigou was describing. They do so by examining both negative and positive cases.

Coase (1960) showed that there is no more certainty that the pollutees have a right to clean air than the factory owner has the right to use the air. Clean air is a common, open access resource owned by neither party. There is a mutuality of interest rather than an automatic onus upon the polluter. Coase's analysis demonstrated that efficiency will be arrived at where exclusive

Table 9.2
Social marginal gain from a factory operation

Output (1)	Marginal private economic gain (5)	Value of uncontracted costs (6)	Marginal social gain (7)
	($)	($)	($)
0	0	0	0
1	14	2	12
2	12	4	8
3	10	6	4
4	8	8	0
5	6	10	−4
6	4	12	−8
7	2	14	−12
8	0	16	−16
9	−2	18	−20

rights are given either to the neighbours or to the factory owner *provided* that there are no costs at arriving at the transaction. There are of course as many complications thrown up by this solution as there are insights provided by it. Importantly, it is uncertain how the neighbours in particular could agree on an optimum level of pollution at specific compensation levels, and how they might ensure the factory's pollution is monitored effectively.

Others have pointed out that mankind begins modifying his environment as soon as his existence is significant. Moreover, rights once seized or acquiesced assume a value. The factory owner's costs are capitalised as rents within his production capabilities and should he on-sell the original factory, the new owner would have had the expectation that any free inputs would continue as such.

The real problem with externalities is their pervasive nature. Most attention is focussed on the adverse externalities—the 'bads' like pollution. However, actions of others also result in unmerited increases in wealth. Such positive externalities occur where, for example, a neighbour maintains a highly attractive garden which raises others' enjoyment of their own property and may even increase its value. Other forms of unmerited increases in property values occur where 'gentrification' of hitherto blighted inner city properties takes place. Similarly, certain skills like those of business economists become more highly prized when a more liberal banking regime allows increased competition in this sector. Much the same may be true of journalists, telephone technicians and airline pilots following relaxations of regulatory arrangements in areas where they are qualified. A road development will change the value of properties around it—possibly reducing the value of those properties which are adversely affected by noise, and increasing the value of those which benefit from greater transportation convenience.

Wherever possible, the agents of change will attempt to garner the maximum rents from the change; but full capture will never be practicable. Indeed, it is the lack of full capture of rents that has been responsible for much of the 'trickle down' of wealth from those generating increases in wealth to the community in general. Even where—as in the case of intellectual properties—new forms of rights have been developed, the full capture of the benefits of inventions is barely conceivable. It would require the inventor to charge each user a separate price based on the user's 'willingness to pay'.

Determining the correct overall level of permitted emissions

The various approaches to emission control which have been discussed are alternatives to 'command and control' approaches. Only a pure Coasian approach makes full use of markets. Both taxes and tradeable emissions require limits to be specified by governments rather than traded off between polluters and pollutees as is the case in true markets.

For all other approaches, the government at a minimum specifies the level of tolerable emissions by examining the costs and benefits. Costs of pollution include those impacting on health and the various sensory perceptions. One way of avoiding imposition of officials' choices in this process is to attempt to measure individuals' subjective values. This involves constructing shadow prices based on willingness to pay. This, the contingent valuation method, uses market research techniques in an attempt to determine appropriate values.

Commonly, questionnaires are devised describing the goods under scrutiny and asking how much respondents would be prepared to pay. The techniques of market research are well known and used extensively both in business and politics.

Some of the difficulties of attempting to assign values in this way are exemplified in a study by Tolley and Randall (1985). Researchers inquiring in the Chicago area about the value of preserving air quality in the Grand Canyon expressed the question in two different ways:

(1) for the Grand Canyon alone, after respondents had been shown photographs; and

(2) as part of a three part sequence which sought values for cleaner air in Chicago, in the Eastern United States and in the Grand Canyon.

In the first study, the value per head for clean air in the Grand Canyon was $90, whilst in the second it was $16. Values were calculated by asking each respondent how much they would be willing to pay for a clear view. Individual valuations were then summed and divided by the number of respondents to give the 'average' benefit each would receive from eliminating air pollution.

The study exemplifies the pitfalls inherent in contingent valuation. The approach's deficiencies are first, it specifies values based upon average utilities whereas demand and supply for goods in general is determined by marginal costs and marginal benefits. Thus a given consumer may value Bounty Bars at $20 and, at a market price of $1, obtain surplus value of $19; but this surplus value is irrelevant to decisions about whether or not the good is produced—or even how much of it is produced. Supply of any good, whether it be clean air or Bounty bars, should continue up to the point where the extra marginal costs of supplying it exactly equal the extra benefits obtained. The consequence of taking a decision based on average costs and average benefits is an all or nothing outcome; but it is more likely that we would trade-off some benefits for some costs.

As suggested by Freeman, a more accurate replication of markets can be constructed using 'choke-off' prices. 'Choke-off' prices are determined by attempting to construct a demand curve of the marginal benefits from pollution reduction. People are asked what they would be prepared to pay for a range of marginal improvements in air quality. Individuals' 'willingness to pay' measures are summed for each marginal improvement. The demand curve generated is a vertical aggregation of individual marginal valuations following the methodology set out by Samuelson (1954). Vertical aggregation is required, in contrast to the normal method under which demand curves are summed horizontally, because pollution is a non-rival good—one enjoyed or suffered by all. The pleasure one person obtains from a clear view (in the absence of congestion) does not stop others from also obtaining satisfaction from it.

The synthetic demand curve may then be used to determine where the marginal costs equal the marginal benefit. However the methodology is still suspect, because respondents are not obliged to make real world choices within the constraints of their budgets.

In determining their choices of goods and services, people face a galaxy of options but are only able to satisfy a limited number of needs. If, added to the air pollution questions in the Tolley and Randall study cited above, respondents were also asked about preservation of wildlife, forests, river purity, parkland in their neighbourhood and the whole host of other facets of life which could be considered externalities, and if people were to be confronted with the real trade-offs between these goods and correspondingly fewer of the goods they would normally purchase—the values would likely be only a tiny fraction of those assignable from seeking answers to single issues.

Freeman expresses a scepticism about the non-use values reported in these studies. He says:

> At issue is whether the responses are measuring a true willingness to pay as defined in our basic theory of individual preferences or whether they are indicators of a general sentiment for environmental protection or preservation that is only imperfectly related to the willingness to commit

resources in a true market or quasimarket setting.

These reservations have undoubted merit. However, survey methods do allow values to be placed on particular resources by those seeking their preservation. And, by placing upper boundaries on the amounts of resources available, more rational choices may be possible.

Approaches to reducing unwanted residuals

Almost all activities impose some cost or benefit on other parties; and, though externalities have been the subject of a lengthy literature, the normal procedure (both for the community as a whole and in economic analysis) has been to neglect them. To a considerable degree this corresponds with efficiency. Arranging and monitoring contracts can be expensive; to attempt to build ledgers of all cross-payments each of us owes and is owed would impose excessive costs and inflexibilities.

This explains one reason why the broad sweep of externalities has received only cursory attention in the past—they have not much mattered. Clean air was abundantly available. Where externalities became important, as in the case of downstream pollution or noise, law developed to take them into account by adapting property rights. Often the increased importance of such externalities, has generated incentives for new techniques to be developed in order to control them better because of asset reductions.

Where unwanted outputs are to be reduced this can be accomplished by:

- lower production of the good of which it is a by-product (this might entail changing the composition of national income so that resources are redirected to outputs having a lower level of deleterious by-products);
- improved efficiency in producing the good so that fewer adverse side effects accompany its output;
- recovery of residual materials and recycling them;
- dilution of the 'bad' so that its effect is less concentrated. As Huber (1985) puts it 'dilution may in fact be a very good

control strategy. As countless cancerous rats might attest, many things are harmful in large concentrations but innocent or even beneficial in small ones'.

Bernstam (1989) demonstrates how the relationship between residuals and output is non-linear and varies over time and between economic systems. Thus in the US, between 1940 and 1970, prior to major efforts to reduce pollution, emissions increased by 30.1% while GNP increased by 212%. From 1970 to 1986, total pollution declined (by one third), in part because of regulatory action. In the main, however, both periods' trends were attributable to shifting composition of national income and technological improvement spurred on by the ceaseless contest of competitive firms to improve their profitability—one means to which is conserving use of material inputs.

Bernstam also shows that this same pattern is not in evidence in the Soviet Union, where emissions of air pollution in 1987 were more than twice those of the US even though national income was probably less than one third that of the US and population only 17% higher.

In the USSR, the economic system is driven by forces other than a profit based meeting of consumer needs at the lowest price. As a result, two factors bring about a considerably higher level of waste and unwanted residues than in comparable market economies.

First, excessive inputs are allocated to capital production due in part, to inefficient machinery. Thus, the share of consumer goods in national expenditure has fallen from 60.5% in 1928 to 24.9% in 1987; and capital investment's share is at least twice that of typical market economies, without growth compensating for this denial of immediate consumption. In short, excessive resource inputs are spent on machines and residual outputs emerge while providing little contribution to well-being.

Secondly, the 'command and control' mechanisms used in the absence of profit related measures must focus upon inputs: the productive unit has its price controlled and its output levels established for it and the only way it can obtain a higher 'profit' is to raise its production costs by requiring increased inputs. In this way the firm is able to pressure the planners into lowering its

Chart 9.6
**Sources of emissions in Sydney —
percentage of total emissions for each pollutant**

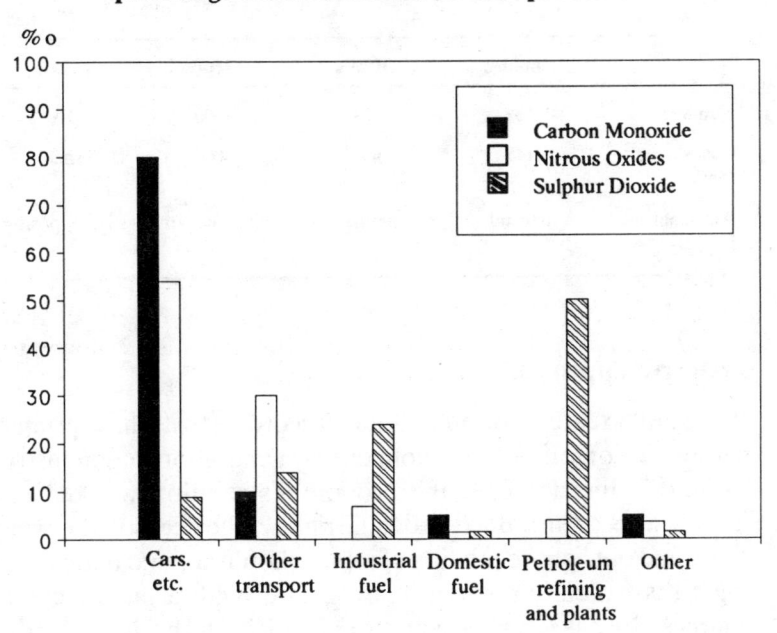

production quotas and raising its output prices.

Use of market instruments to combat air pollution combines the power of individual self-interest with the best sources of information on how to reduce emissions to the levels sought at the lowest cost. The firms and individuals who produce the emissions have the knowledge on how to reduce their outputs most economically. They will seek to take opportunities for gain (or reducing their losses). In doing so, they ensure a more frugal use of resources in meeting standards than would be possible for regulatory authorities, whose information on economical means of reducing pollution cannot be as complete. Because of their vested interest and operational familiarity, firms are likely to be much better informed about the available techniques for reducing residues in the most cost effective manner than are government officials.

Table 9.3
Estimated effects of emission trading 1979–1985

	Bubbles	Offsets	Netting	Banking
Number	132	1000	8000	100
Cost saving ($m)	145	n.a.	4000	small
Air quality impact	neutral	neutral	slightly negative	slightly positive

Preferred approaches

It is useful to categorise air pollution according to its three primary causes: automotive, household energy generation and industrial facilities. In each case the economist's solution would be to impose a tax or introduce tradeable rights. The decisions between these and outright regulation of inputs should depend upon policing costs. Clearly such costs are greater with a multiplicity of sources. Just as it makes sense for electricity authorities to strike separate deals with major users but charge generally available rates to domestic consumers, so it is appropriate for pollution controls to be tailored differently for minor as opposed to major source emission locations. For the former, transaction costs of monitoring market based approaches may be prohibitive given current technological capabilities.

Automotive and domestic sources are far more important than industrial sources, as Chart 9.6 shows with respect to Sydney (which is typical of other Australian cities).

Industrial pollution

In the case of industrial pollution, economies are available if trade of pollutants between different sources is allowed. Crandell (1983), for example, found that the cost of controlling emissions from paper mills was three-fold that of controlling similar ones from metal working factories—fewer pollutants could be achieved at the same social cost by concentrating on the latter.

Industrial pollutant trading can take a number of forms, including netting, offsets, bubbles and banking. Netting sets emission standards for one business but allows trading within plants. Offsets allow new pollutant sources if compensatory reductions can be obtained from other sources. Bubbles are defined for a particular area and allow different pollutants within a given aggregate limit. Banking enables credits to be earned for over-performance and subsequently used or traded.

Hahn and Hester (1987) explain how the EPA has allowed:

- netting since 1974;
- offsets since 1976;
- bubbles since 1981;
- and banking.

Their estimates of the effect of these measures, 1979 through 1985, are set out in Table 9.3.

They consider the effects to have been less than satisfactory because of arrangements by environmental groups which have thwarted some proposals and because of uncertainty by firms. They attribute this latter effect to the need for EPA approval to be specifically granted and the discretion EPA has (and is thought likely to use especially with regard to bankable emissions). Moreover, since 1986 the EPA has insisted that bubble trades be 'taxed' so that there is a net reduction in emissions of 20%.

These problems notwithstanding, the notion of emission trading as a cost effective interventionary tool is gaining increased currency. Congress has agreed to a scheme which splits the US into two areas with unlimited trading of SO_2 emissions permitted within each of them.

Hahn and Hester (1989) also estimate there to be 132 'bubbles' within which trading takes place, and these brought a cost savings of $435 million between 1979 and 1985 while having had a neutral effect on pollution levels. These savings are in addition to savings of up to $12 billion estimated to have accrued from 'netting'—allowing firms flexibility to over-perform in some areas of a plant to compensate for under-performance in others.

Levin (1985) quotes some specific examples where these approaches have resulted in gains. Dupont, facing a requirement to reduce emissions by 85% in each of 119 stacks, negotiated to reduce 99% of emissions in seven stacks which proved faster and over-achieved the aggregate goal at a saving of $12 million in capital cost and $4 million a year in operating costs. General Electric was allowed to forgo $1.5 million in capital expenditure and $300,000 in operating costs, required to meet emission controls in Louisville, by negotiating with International Harvester which was able to over-perform mandatory requirements relatively cheaply.

The Pigovian approach, which for long had been preferred by economists, is to apply pollution taxes. Like trading, this allows greater flexibility for firms in designating the appropriate means of meeting output levels. It also allows compensation of the community at large for residuals in excess of those considered appropriate.

Buchanan (1988) sets the strict conditions under which an externality can legitimately be countered by the imposition of a tax. He suggests that:

- all persons must be equally damaged;
- all must be consumers/buyers purchasing in equal quantities;
- the revenue must be equally shared.

In such rare cases, he suggests, the price would rise so that the 'bad' would be economised upon.

Buchanan maintains correctly that without his strict conditions for governmental action to combat an externality there will be distributional consequences—consequences which will be determined by the political market, and generate social costs via lobbying and government failure. Some redistribution is inevitable where any departure from equal usage and production occurs.

Others, for example Terkla (1984), take the view that effluent taxes improve welfare because they charge the polluter the true economic cost, and allow the replacement of other taxes which are designed to raise revenue and unintentionally distort

Table 9.4
Empirical studies of air pollution control

Study	Pollutants Covered	Geographic Area	CAC Benchmark	Ratio of CAC Cost to Least Cost
Atkinson and Lewis	Particulates	St Louis	SIP regulations	6.00[a]
Roach et al	Sulphur dioxide	Four corners in Utah	SIP regulations Colorado, Arizona, and New Mexico	4.25
Hahn and Noll	Sulphates standards	Los Angeles	California emission	1.07
Krupnick	Nitrogen dioxide regulations	Baltimore	Proposed RACT	5.96[b]
Seskin et al.[c]	Nitrogen dioxide regulations	Baltimore	Proposed RACT	14.40[b]
McGartland	Particulates	Baltimore	SIP regulations	4.18
Spofford	Sulphur Dioxide	Lower Delaware Valley	Uniform percentage regulations	1.78
	Particulates	Lower Delaware Valley	Uniform percentage regulations	22.00
Harrison	Airport noise	United States	Mandatory retrofit	1.72[c]
Maloney and Yandle	Hydrocarbons DuPont plants	All domestic reduction	Uniform percentage	4.15[d]
Palmer et al.	CFC emissions from non-aerosol applications	United States standards	Proposed emission	1.96

Notes: CAC = command and control, the traditional regulatory approach.
SIP = state implementation plan.
RACT = reasonably available control technologies, a set of standards imposed on existing sources in non-atttainment areas.

[a] Based on a 40 $\mu g/m^3$ at worst receptor.
[b] Based on a short-term, one-hour average of 250 $\mu g/m^3$.
[c] Because it is a benefit-cost study instead of a cost-effectiveness study, the Harrison comparison of the command-and-control approach with the least-cost allocation involves different benefit levels. Specifically, the benefit levels associated with the least-cost allocation are only 82% of those associated with the command-and-control allocation. To produce cost estimates based on more comparable benefits, as a first approximation, the least-cost allocation was divided by 0.82 and the resulting number was compared with the command-and-control cost.

economic choice and resource allocation. Terkla suggests that the total revenue raised from an efficient tax, which in 1982 he estimated would optimally be set to raise between $1.8 and $8.7 billion, would generate considerable efficiency gains. Based on Browning's estimates of the welfare losses generated by levying income tax, $0.35 per dollar collected, he estimated effluent taxes would raise welfare by $630 million – $3.05 billion.

Terkla's effluent tax is therefore seen as more efficient than the alternative means of raising revenue. Like others, he sees merits of such a tax system in allowing the market to discover the most efficient means of adapting to a new incentive structure. His estimates are based on effluent taxes not generating the sort of losses from work/leisure substitution which Browning's estimates are projected. Hence, although many would dispute the predicted cost effectiveness, there is wide agreement that making use of the market in this way will generate economies.

In practice, taxation of residuals has not found favour. Those facing the taxes have of course objected whilst environmental activists have opposed this policy approach because of the apparent endorsement it implies to the generation of pollution.

There are in addition several practical difficulties in devising a workable taxation regime. Buchanan (1988) points to one major difficulty. The various parties are likely to have different interests. Those producing, or using most intensively, the output of the polluting facilities are likely to wish to see the tax levied at as low a rate as possible; those on whom the impacts of the residues fall most heavily are likely to favour a prohibitive level of tax; taxpayers who are relatively indifferent to the pollution and the output the facilities' produce are likely to favour a tax which maximises the revenue raised so that other taxes might be reduced. How are these differences to be reconciled? The obvious gains will lead the parties to engage in wasteful lobbying exercises to promote their particular interests. Can we be confident that governments will arbitrate dispassionately?

Pollution taxes are based upon the fundamental principle that the polluter pays. Yet this tends to treat the polluter as the malefactor and the pollutee as the victim. In fact there is a mutuality of interest. The unowned resource did not belong to the pollutee in the first instance. As soon as mankind breathes air some impression is made on the natural environment. It is no more certain that the population surrounding a polluting factory has the rights to clean air than the owners of the factory have the right to soil it. This statement is even more graphic in situations where the factory owner was there first and the 'victims' moved in later (perhaps to take advantage of the opportunities to improve their well-being offered by locating close to the factory).

Industrial air pollution is more easily combated by granting tradeable rights to pollute. Means of monitoring major sources of pollution are readily available as is the technology to effect this. Hartley and Porter (1990) draw attention to the application of deuterated methane, a chemical tracer which mimics SO_2 to detect sources of pollution in southern Utah.

Tietenberg (1990) assembles eleven empirical studies of market based approaches to pollution control compared with the 'command and control' approach. Each study estimates the cost of a 'command and control' strategy limiting pollutants in a localised area. These approaches are contrasted with the least cost allocative method involving either trading or taxes. Table 9.4 gives the ratios of the 'command and control' outcome to the estimated least cost allocative mechanism.

As Tietenberg points out, the estimated savings are theoretical—they are gains achievable on the basis that sunk costs have not been incurred, perfect information is available and multilateral trades take place. Moreover, if emission credits are traded on a pollutant-by-pollutant basis, rather than on an amalgam of pollutants, the trades themselves are rendered considerably more complex. Nonetheless, the wide number of studies, (each of which demonstrates considerable gains from applying market principles), present powerful evidence against 'command and control' methods.

There is little use made of economic instruments to control Australian industrial emissions. This may change. Both the NSW Government and the Commonwealth Treasury have placed on record their favouring of market based measures where these are possible. Indeed, the New South Wales Government has announced its intention to place a greater priority on 'pollution taxes and charges, pricing of services based on true costs, tradeable emission rights and government subsidies' (Greiner, 1990).

At the present time, however, pollution in Australia is combated only by 'command and control' methods. Some flexibility is provided for in certain circumstances. Thus, the Victorian Government, in introducing more stringent requirements for the control of conveyor equipped coating lines in 1988, specifies input controls in detail but allows firms to meet the standard in other ways providing they are able to demonstrate that these achieve equivalent results. Even in these cases however, the regulations contain rigidities over and above the absence of provision for trading. These include grandfathering provisions which discourage the replacement of equipment.

Automotive and households

It has previously been suggested that opportunities to make use of market mechanisms are limited. Monitoring and other transaction costs may make even these partial market type approaches inapplicable for determining efficient household and vehicle pollution behaviour.

In the case of motor vehicles, governments the world over have introduced standards for emissions, more recently by seeking the use of lead-free petrol. A general approach may be the rational solution, notwithstanding that in Australia, citizens of places like Albury Wodonga with little pollution would not obtain value from the increased capital and operating costs involved (the latter partly hidden by the governmental requirement that lead free petrol be cross subsidised by leaded petrol)[1].

The move to lead free petrol has contributed to a marked reduction in the level of lead in urban areas. Thus both in the centre of Melbourne (where it was previously at double the level set as acceptable), and in the suburbs, lead levels have exhibited a considerable decline.

Grenning (1985) is critical of Australian mandatory emission control standards adopted in Australia Design Rule 37 (ADR 37) during 1986. He favours emission charges over the technical solutions introduced. Grenning maintains that the standards adopted were overkill because:

- any problem which occurs is confined to Sydney and to a lesser degree Melbourne (which together on the widest interpretation, might account for 30% of the vehicle population); and
- pollution levels in these cities had began to decline anyway as a result of industry restructuring and relocation.

ADR 37 meant a cost per vehicle, at 1985 prices, of $70–160 which is in addition to a slightly higher impost introduced by the previous standard. Important shortcomings of a standard like this are that they apply only to new vehicles—and perhaps therefore to only 12% of the annual stock. In addition, the increased cost (and reduced performance) creates disincentives to replace existing vehicles and therefore, to some extent at least, has perverse effects. Furthermore, achieving the targeted output of emissions by using a 'command and control' approach is far from certain. It depends crucially upon the vehicles being properly maintained and is totally negated if owners disconnect the control mechanism.

Grenning favours a charge based on the outcome of emissions as measured at the annual vehicle test. Although making use of more direct and effective controls, this would also have shortcomings:

- the annual inspection is only a once per year measure and there will be ways discovered which would allow vehicles to demonstrate a short term measured acceptability in emissions;
- it may entail greater costs if owners are required to have modifications undertaken retrospectively;
- it entails some administrative costs, and if cars can be registered in areas where these additional costs are not required there will be considerable incentives for evasion.

Leaving aside the issue of whether or not mitigation of emissions specified for Australia was necessary, it is not clear that the 'command and control' solution is inferior to the generally preferred output based solution in this particular case.

In the case of households, many locales have banned coal and wood burning, though the latter somewhat ironically has more recently shown an increase in popularity because it is thought to be a more natural fuel. Prohibition would not be the preferred solution of most economists yet it may well be more efficient than imposing an easily evadable tax. As with garbage, it is not easy to see how contractual difficulties can be overcome to allow tradeable rights to operate with respect to this source of air pollution.

Rural pollution

Aside from the issue of urban air pollution—one which largely involves health and unpleasant smells—there are issues of rural air pollution. In the main these involve maintaining a pristine air quality. Such issues have not assumed any importance in Australia to date and major industrial sources of air pollution facilities in areas of high natural value are most unlikely to be economically justified. A contemporary exception is the controversy over the location of a high temperature incinerator to service the southeast corner of the continent.

Environmental regulatory costs

It is often pointed out that ecology and economics have much in common in so far as both start from the premise that everything is interconnected. Many point to a comity of interests between the two frameworks. Some, like Hamrin (1981) suggest that the application of environmental standards on emissions required by government regulations will actually benefit both the environment and the economy by saving energy, virgin resources and so on. Such assessments glide over the costs in terms of resources required and income foregone in achieving the regulatory standards.

Others, more conventionally, suggest that such a comity exists, and would be the natural outcome of market forces if property ownership rights could be adequately defined to prevent excessive use of 'unowned' resources and the consequent externalities generated. Fred Smith (1989) goes further than this and maintains that modern technology can allow all externalities to be internalised.

The mainstream view is that for some goods the externality looms so large and the difficulties of internalising it are so great that interventions by government are essential. Such interventions can only be legitimate where they are based upon the construction of shadow markets for evaluating the worth of those activities where externalities inhibit provision by natural markets. Although there have been no analyses of the aggregate costs of environmental regulation in Australia, a number of studies of the costs of air and water pollution control have been conducted in the US. These include many estimates of the costs of environmental protection on economic growth by Crandell; Christainson and Haveman; and Conrad and Morrison. But perhaps the most rigorous has been that of Hazilla and Kopp (1989). Taking the Environmental Protection Agency's (EPA) cost estimates of federally mandated pollution controls ($425 billion in 1981 dollars), $648 billion in 1981–1990 current dollars), the authors apply elasticities of substitution, both to the economy's outputs and inputs. Because of substitution, the initial estimates are lower than those derived from EPA's engineering based estimates. Both consumers and producers take actions to alleviate the cost burden which regulation imposes, for example, by switching purchases to goods which do not have additional cost requirements. In this way the aggregate cost imposition is muted. But the dynamic, secondary impacts of these costs must also be factored in. In addition, the effects of the intervention cannot be confined to one particular time period but will flow on into subsequent periods. The resulting costs from making these adjustments are estimated at $977 billion, which by 1990 translates into a diminution of real GNP of 5.9% and of investment by 8.4%.

This approach, however, tends to take preferences and technical capabilities as given. In fact, both consumers and producers can make rapid adjustments. And over time alternative needs and new means of meeting them are found while both supply and demand curves for a particular good tend to become flatter.

In the case of consumers, for example, the introduction on congested bridges of express lanes which may only be used by multiple occupancy cars brought considerable behavioural changes both in Sydney and in San Francisco. Consumer adjustments to picking up or accepting rides from total strangers were remarkably swift and it is difficult to argue that the costs estimated at the outset prevail in anything like their original magnitude after a short transition period. On a larger scale, adjustments following the implementation of major infrastructural changes within cities—changes which were envisaged to affect property values markedly—have been absorbed without lasting declines in these values. For example, converting vehicular roads within cities to pedestrian malls has often brought very rapid behavioural changes on the part of shoppers which were unanticipated and which totally negated the adverse effects previously expected.

For producers, the very rapid adjustment of some industries to the four-fold increase in oil prices, which took place in the 1970s, reveals great flexibility. The Japanese steel industry converted from oil firing to making more efficient use of coal, a formidable energy-saving innovation. Entrepreneurial reactions like this cannot be incorporated into general equilibrium models except by using non-scientific 'fudge factors'. Indeed, the inability to account for the role of the entrepreneur in seeking out opportunities constitutes perhaps the greatest shortcoming of all economic modelling.

Because of compensatory variation, it is probable that the economic analysis of environmental quality regulation overstates the costs to society.

There are however other factors which would tend to operate in the opposite direction. One is that entrepreneurship itself is a scarce resource and energies directed at ameliorating an intervention might be energies which would otherwise be directed at discovering new means of adding value. In addition, modelling typically assumes zero transaction costs, perfect factor mobility and other notional attributes which we tend to group under the heading of perfect markets.

The work of Hazilla and Kopp, originally commissioned by the EPA, has a strong following within the agency, even though its findings have not been formally endorsed. Publicly the EPA quotes a more conservative, less comprehensive cost of environmental interventions which amounts to only 1.7% of GNP. Nonetheless, the Hazilla and Kopp work constitutes the state of the art in estimating environmental costs and, because it examines the total picture, is superior to those estimates which confine their impacts to specific sectors.

The costs of environmental regulation estimated by Hazilla and Kopp incorporate only the costs of those regulations falling under the control of the US Environmental Protection Agency. These cover air and water pollution and waste disposal. They do not include other regulatory interventions which fall within the environmental embrace, such as use of forests and wilderness, protection of flora and fauna and measures to combat soil erosion.

Concluding comments

The magnitude of the costs involved in effecting pollution control makes the means by which this is undertaken of considerable importance to general well-being and not least to industrial competitiveness. Compared with other countries, Australian levels of pollution are low—in part because our cities tend to have fewer industrial facilities and their populations are less concentrated.

In addition, pollution levels have been reduced, notwithstanding industrial growth and far greater numbers of automobiles. This outcome, welcome as it is, is the result of 'command and control' policies. Such approaches may well be unavoidable and the most effective means of combating the minor source domestic and automobile emissions which together account for the preponderance of urban pollution. They have been demonstrated not to offer the lowest cost strategies for control of emissions from major industrial sources. Australian authorities, however, have not sought to apply market-based solutions, except in limited cases where offsets within major sources have been negotiated.

Endnote

1   Interestingly, however, a survey about environmental concern conducted by the Australian Bureau of Statistics (Cat. No. 4115.0) listed concern about pollution as being highest in the two Australian regions, the Australian Capital Territory and the Northern Territory, where problems in this

regard would be much less evident than elsewhere. This may reflect the preferences of people living in those two regions. It may also reflect a heightened awareness about environmental matters generally, a possibility which is perhaps corroborated by generally enhanced levels of concern registered about other environmental issues including, nuclear power, nature conservation, old growth and rain forests, soil erosion, and water salinity, etc.

10 The enhanced greenhouse effect

Executive summary

Theories that increased emissions of gases attributable to mankind's activities are causing global warming have rapidly assumed major prominence. As the upper atmosphere cannot be owned there are no mechanisms stemming from individual property rights which would allow its use to be properly valued. There are therefore no automatic mechanisms to allow careful stewardship of the resource. Hence there are concerns that any action to combat global warming may be paralysed by an inability to exercise controls.

It is certain that there has been a build up of carbon dioxide, and the CFC family of gases since industrialisation and the growth in human populations began to accelerate. Nonetheless, the facts are obscure. Not the least of these is whether or not, despite the increase in gases with a capacity to retain terrestrial heat, and whether significant warming will actually take place. There are offsetting phenomena like ocean absorption and cloud formation which can easily counteract the estimated effects. Further, if global warming were to take place, it is uncertain whether or not it would bring net harmful effects. However, many countries, including Australia, have already adopted some limited commitment to reducing their emissions of greenhouse gases and there is pressure to take further steps.

This chapter examines some of the evidence concerning the issue. It draws attention both to deficiencies in the theory which seeks to explain the phenomenon, and analyses the evidence that warming might be taking place.

The chapter goes on to outline procedures which might be used to reduce greenhouse effects should this prove to be a wise course of action. It draws attention to market-based instruments, like tradeable rights to emit greenhouse gases and taxation measures, as being considerably more efficient than governmental restrictions and requirements to use particular technological solutions. Market-based approaches enlist greater knowledge about potentially cheaper means of achieving reduced emissions and incorporate a flexibility which is absent in heavy handed command and control measures.

Introduction

Global warming has taken centre stage among environmental issues of the 1990s. Average global temperatures are thought to be rising because of increased concentrations of greenhouse gases. These gases act as a blanket around the earth modifying its energy balance. Over the last 200 years, man's activities have raised concentrations of the naturally occurring gases, CO_2, methane, and nitrous oxide, bringing about higher levels of retained heat. New man-made gases, with a similar potential effect—such as chlorofluorocarbons (CFCs) and the halon class of gases—have also been released into the atmosphere. CFCs and halons are also thought to be responsible for depletion of upper atmospheric ozone levels.

If it is occurring, global warming bears all the marks of a classic environmental problem. No-one owns the upper atmosphere, and the individuals, industries, flora and fauna discharging gases into the atmosphere pay no price for the privilege. With open access to all, Hardin's 'The Tragedy of the Commons' seems inevitable (Hardin, 1968). Each individual source of greenhouse gas generates a benefit larger than its private costs. In contrast, the costs each source avoids incurring directly are thought to be large and widely spread. Presaged on these assumptions, there are calls for an internationally coordinated programme of greenhouse gas reductions.

In 1988, a conference in Toronto, with delegates from 46 countries, suggested setting a target of a 20% reduction in emissions of the major greenhouse gas, carbon dioxide, by the year 2005. There is now a growing group of nations who have responded to this suggestion and moved towards setting greenhouse targets. However, only Sweden has given legislative commitment to a target. West Germany is aiming at a 25% cut. Denmark, East Germany, New Zealand, Austria and Italy are moving towards 20% targets. At least nine other countries, including Australia, Canada, the UK and Japan, have indicated some commitment to stopping the growth of greenhouse gases.

There are three dimensions to the greenhouse gas and global warming debate:

- determining the facts;
- measuring the effect;
- setting out policy options.

Determining the Facts

While large increases in greenhouse gas emissions have indisputably occurred since the industrial revolution, the evidence for any resulting global warming is at best scanty and at worst contradictory. Theoretically, there is a case for expecting global warming but conclusive empirical evidence to support the hypothesis has not been found. Without such evidence the phenomena remains a theory, no more plausible than the impending ice age which was being predicted by some climatologists in the 1970s.

Measuring the Effects

Assuming significant global warming has or will occur, it could have both positive and negative effects. Emphasis has been placed on the negative effects—higher average, global temperatures and rising sea levels; but there are also positive effects, including longer growing seasons and increased precipitation. In addition, increased concentrations of carbon dioxide can be beneficial through accelerating plant growth and lowering water requirements for crops. If the greenhouse phenomena proves well founded, both positive and negative implications must be assessed before deciding that it poses a potential problem.

Policy Options

The difficulties associated with defining rights to the global atmosphere suggest some form of government intervention could be required to overcome any global warming problem. These policy prescriptions could be either adaptive or pre-emptive. If the problem is expected to be large and the effects certain, the preferred approach is likely to be some form of pre-emptive action to reduce emissions of the greenhouse gases. The instruments to achieve this could include carbon taxes or tradeable emission quotas. However, if the effects of the problem are relatively minor and the costs of taking early action to mitigate these are high, a more adaptive approach would be preferable, with governments and markets responding to the effects as they become apparent.

Policy decisions should be strongly influenced by two guidelines. First, decisions should be based on measurements of the costs and benefits of *different* levels of increased greenhouse gas (GHG) emissions. GHGs are the result of productive processes which yield a stream of income, thereby enhancing the quality of life. Consideration of a freeze or cuts in GHG emissions must take these benefits into account. Logically, the preferred response should be to allow additional increases in GHGs up to the level where the potential costs outweigh any gains. Policy should not generally be polarised into all or nothing choices. A full range of incremental options should be considered.

A second guideline for policy is that, where possible, market mechanisms should be used in preference to legislating specific technologies and productive processes. If a pre-emptive policy stance is to be taken, establishing tradeable quotas or, failing that, carbon taxes is preferable to requiring specific energy efficient or carbon dioxide minimising technologies. Allowing the market to determine the means of achieving some national standard gives industry the flexibility to choose the most cost effective methods. Assigning tradeable quotas to set levels of emissions would allow firms which are able to reduce their emissions at least cost to over achieve and be compensated by firms which find reductions more expensive. A carbon tax, set at a rate to achieve the same level of emissions, would also allow flexibility in technology choice, (although not the additional cost saving to be found when firms may trade emissions).

A number of more market oriented policies, which are sensible in their own right, could also reduce GHG emissions. For example, the implementation of efficient pricing mechanisms by state owned electricity corporations may, by itself, also go some way towards achieving lower GHG emissions.

Implications of a global environmental problem

The global nature of the greenhouse phenomena distinguishes it from other environmental and resource related problems discussed in this book. If global warming is a problem, its solution would seem to require very wide cooperation. Large and certain costs are a high price to pay for benefits which, for any single emitting nation, would be shared among all and would most certainly be small and highly uncertain. Hence, the incentives for an individual nation to renege on internationally agreed targets are high. As a result of this, in the past, wide and costly actions by sovereign states in pursuit of goals with joint pay-offs have proven difficult to achieve. However, confidence in the possibilities of global cooperation and multi-lateral treaties to counteract global climate change received considerable impetus with the success of the 1987 Montreal Protocol on Substances that Deplete the Ozone Layer. The Montreal Protocol has been ratified by many countries, including Australia (in 1989). It requires phasing out of CFCs—used in a wide range of applications including refrigeration and aerosol propellants.

The success of the Montreal Protocol has encouraged the call for an international treaty imposing global limits on GHGs. However, any such treaty will have far larger ramifications than the phasing out of CFCs. Global limits on GHGs would have major and direct implications for the energy and transport sectors of national economies. Both these sectors are major sources of carbon dioxide emissions, which is the GHG causing most concern. All sectors would be indirectly affected because of the widespread linkages the energy and transport sectors have to the rest of the economy. Although pricing action since 1973 has considerably slowed the growth of energy demands within developed economies, that demand has not contracted. Moreover, rapid growth in energy demand continues to be experienced in the more successful developing countries. The fastest growing energy sectors in the world are in the Asian region with a growth rate of 6% in primary energy consumption in 1989 (Foster 1990). Most of this growth has and will continue to come from coal fired electricity generation.

If global limits on carbon dioxide emissions are to be contemplated, conflicting views are likely to be held on the appropriate basis of these limits. If based on existing usage levels, global limits on carbon dioxide emissions would have serious implications for the economic growth of the world's poorer nations. If quotas were to be allocated on, say, a per capita basis this would result in income redistribution to those nations. By tentatively agreeing to *reductions* in emissions, the developed nations are registering their opposition to the issue becoming instrumental in income redistribution. These wide reaching implications warrant careful consideration of proposals for limiting GHGs.

In the rest of this chapter, we first discuss evidence for the enhanced greenhouse, global warming phenomenon. Secondly, the likely effects of this hypothetical scenario are discussed. We then present some market based policy prescriptions for counteracting potential problems. Finally, we examine the likelihood and desirability of achieving international agreement on GHG targets.

The evidence for greenhouse gases and global warming

The mechanisms involved

Since the 1890s, there have been scientists who have expressed concern that increases in greenhouse gases related to human activities will enhance the greenhouse effect raising global temperatures further. Strictly speaking it is this enhanced effect which is again being debated.

A knowledge of how the earth absorbs and radiates energy is essential for understanding the greenhouse effect. Briefly, the earth receives radiated energy from the sun mainly in the visible band of the electromagnetic wave spectrum. In turn, this energy is re-radiated out into space by the earth, mostly at night. Because the earth is much cooler than the sun, the re-radiation occurs at much longer wavelengths in the infra-red range of the spectrum. Water vapour and greenhouse gases, such as carbon dioxide, act as a 'blanket' around the earth reflecting some of this infra-red radiation back to the earth's surface. Consequently, temperatures at the earth's surface must rise until the amount of radiation intercepted from the sun, and re-radiated back to the earth by greenhouse gases, is equal to the amount of infra-red radiation escaping into space.

This temperature enhancing effect can easily be observed by

Table 10.1
Australian sources of greenhouse gases and their contribution to the greenhouse effect

Greenhouse gas[a]	Source	Emission (million tonnes per year)	Relative contribution to greenhouse effect (%)
Carbon dioxide	Coal	158	27
	Oil	72	12
	Gas	25	5
	Other (Cement, natural gas flares)	12	2
	Total	**267**	**46**
Methane	Cattle, sheep etc.	2.2	6
	Bushfires	2.0	6
	Garbage tips	1.5	5
	Coal mining/handling	0.3	1
	Natural gas leaks	0.2	1
	Rice	0.2	1
	Other animals	0.2	
	Total	**6.6**	**19**
Nitrous oxide	Agriculture	0.3	17
Chlorofluro-carbons	Refrigeration, aerosols, etc.	0.012	18

[a] Ozone, although a greenhouse gas, is not included as it is only emitted in insignificant amounts in the Southern Hemisphere

comparing temperatures on a cloudy and a clear night at the same time of the year. The phenomena is also analogous to the warmer temperatures experienced in a greenhouse, and hence the descriptive name, greenhouse effect.

The major greenhouse gases released by human activities are CO_2, CFCs, methane and nitrous oxide. (Water vapour from natural sources is in fact more potent than any of these gases in maintaining warmth on the earth's surface.) Different gases absorb infra-red radiation at different wavelengths and vary in their potential effectiveness in creating a greenhouse effect. Though carbon dioxide concentrations are many times higher than methane and CFCs, both the latter gases contribute far more to the greenhouse effect per unit of concentration. Compared with CO_2, methane is 36 times more effective; CFCs are 14500 to 15000 times more effective, though concentrations are extremely low. Effectiveness is also related to the length of time a gas remains in the atmosphere. Table 10.1 shows the relative contributions various Australian sources of greenhouse gases could make to the greenhouse effect after allowing for differences in warming capabilities and length of time remaining in the atmosphere. These differences in effectiveness have important implications when considering policy alternatives for counteracting any potential increase in the greenhouse effect.

Greenhouse gases are an essential ingredient for maintaining life on earth. They raise global temperatures by about 33 degrees Celsius (Gribbin, 1988; Landsberg, 1989)—just enough to make life comfortable. Without the blanket created by greenhouse gases, average temperatures on the earth's surface would be −18 degrees Celsius. Just above the surface of the earth, average temperatures are 15 degrees Celsius. There is solid evidence to show that greenhouse gases have increased since the time of the industrial revolution, 200 years ago. But, though there is a theoretical basis for expecting a corresponding increase in global temperatures, it is questionable how large and how important this increase in temperatures may be and the empirical evidence for global warming is inconclusive.

Increased atmospheric gas concentrations of greenhouse gases

Chart 10.1 shows the increases in concentrations of the major GHGs since 1750. Since pre-industrial times, atmospheric CO_2 has increased by 26% to a level comparable with that of 150,000 years ago; atmospheric methane and CFCs have increased markedly from negligible levels. Much of the increase in carbon dioxide concentrations can be attributed to the burning of coal and other fossil fuels. At one time the carbon in fossil fuels was itself a part of biological systems, and would have passed from the atmosphere to the biosphere and back again through the processes of photosynthesis and respiration. Fuel combustion completes another link in this vast carbon cycle—reconstituting carbon as atmospheric carbon dioxide and making it available to

Chart 10.1

Concentrations of carbon dioxide and methane after remaining relatively constant up to the 18th century, have risen sharply since then due to man's activities. Concentrations of nitrous oxide have increased since the mid-18th century, especially in the last few decades. CFCs were not present in the atmosphere before the 1930s.
Source: I.P.C.C. (1990)

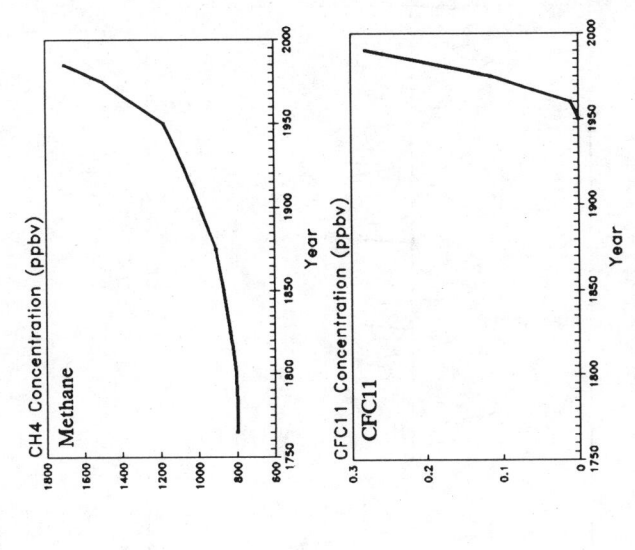

Chart 10.2
Global mean combined land-air and sea-surface temperatures, 1861–1989, relative to the average for 1951–80

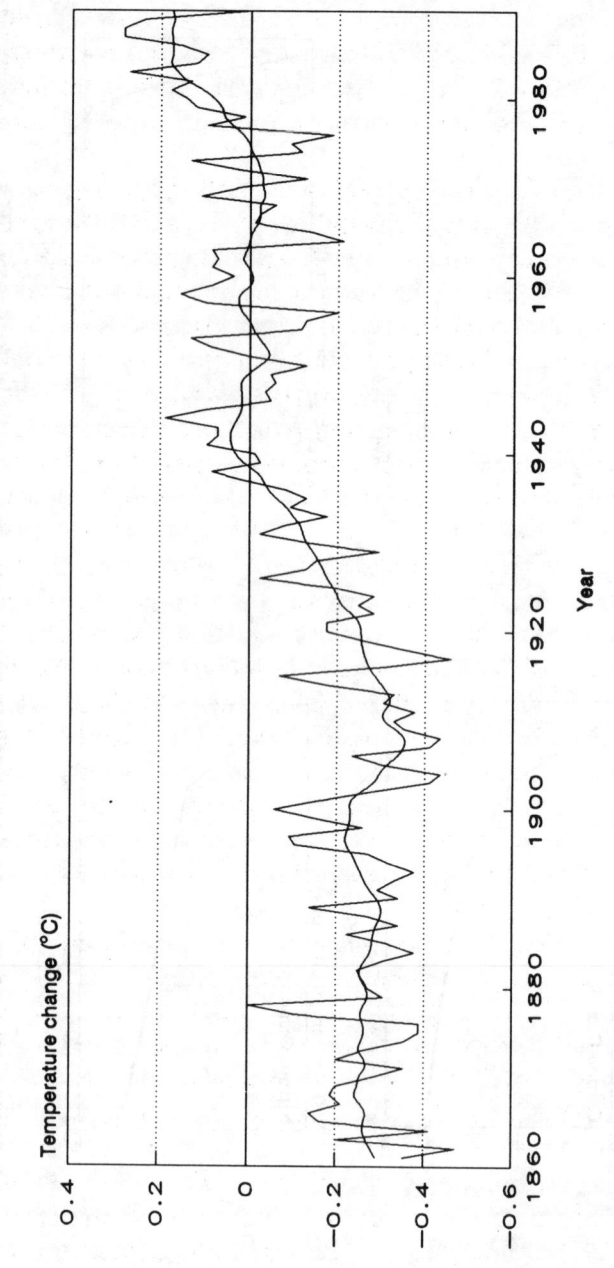

Source: I.P.C.C. (1990)

plants once again. Deforestation also affects the cycle by reducing another 'sink' in which carbon is stored and decreasing the area of forest also available for absorbing—or more correctly 'fixing'—atmospheric carbon. Consequently, reductions in forest area have also contributed to higher carbon dioxide concentrations in the atmosphere.

Whereas carbon dioxide is a naturally occurring gas which is constantly reconstituted in the biosphere, atmospheric CFCs are synthesised compounds directly attributable to man's activities. For the most part, their synthesised compounds remain resident much longer than naturally occurring gases.

The causes of increased concentrations of methane and nitrous oxide are less certain than for CO_2 and CFCs. Increases in concentration of methane could be related to increased livestock populations, and emissions from garbage tips and rice paddies. Nitrous oxide increases are also probably related, in the main, to agricultural activity.

A number of empirical sources provide consistent and well documented evidence of increases in atmospheric gas concentrations. Carbon dioxide concentrations measured in air bubbles trapped in glacial and Antarctic ice show the concentrations were stable at around 280 parts per million (ppm) for thousands of years prior to the industrial revolution. During the eighteenth century, concentrations began to increase and are now around 350 ppm (Landsberg, 1989). Similar measurements of methane concentrations in ice air bubbles have shown an increase, over the last 300 years, almost exactly corresponding to growth in human population (Pearce, 1989).

The first accurate atmospheric measurements of carbon dioxide were taken at Mauna Loa (Hawaii) and the South Pole in 1957–58. As these locations are far away from industrial pollution sources, they provide a good measure of the 'mixed' state of the air. (Gribbin, 1988). Similar measurements have also been made in Australia since the late 1970s for carbon dioxide, methane, chlorofluorocarbons and nitrous oxide. Consistent increases in CO_2 have been measured over south eastern Australia. Even more pronounced increases in methane, CFCs and nitrous oxides have been measured at the Cape Grim Baseline Observatory in north-west Tasmania (Landsberg, 1989)

The evidence for increases in temperature

Evidence of temperature changes over the past 140 years does not confirm the relationship predicted between GHG concentrations and global temperatures. Chart 10.2 reproduces the most commonly cited historical time series data for temperature change. While this data shows an upward trend since the 1860s, it is an inadequate reflection of the predicted relationship and the increase is far less rapid than would be expected. Most of the increase in temperature occurred between 1900 and 1940, when GHG concentrations were increasing more slowly than at present. Moreover, between 1940 and the early 1970s average temperatures fell. Comparing the data from Charts 10.1 and 10.2, it can be seen that gas concentrations increased sharply during that period. The lack of a direct correspondence between temperature and gas concentrations has lead scientists to suspect that other factors have generated the changes in average global temperatures (Mason, 1989).

These doubts are strengthened by the sparsity and non-random nature of the sample data locations. The whole of the Atlantic ocean is represented by four island stations, and the majority of measurements have been taken in cities—which are heat islands. In addition, the series begins in 1880, the year when records were standardised. However, 1880 is also thought to have been a year in a unusually cold period. Lindzen argues that if years prior to 1880 were included overall change would be '0.1 degree plus or minus 0.3 degrees Celsius' (Beckman, 1989, citing a lecture by Richard Lindzen, Professor of Meteorology at MIT). The variability of the data series is also greater than any long term trend, which raises further questions as to the statistical validity of assuming such a trend exists.

All these factors explain the conflicting reports about global warming. Basing their statements on time series similar to the one shown in Chart 10.1, some scientists have labelled the 1980s 'the hottest decade since records began'. These claims were widely reported in the newspapers in early 1990, and credence was lent to them by the United Nations Environment Program's Intergovernmental Panel on Climate Change (IPCC, 1990). However, in February 1990 it was claimed that, after correcting for heat island effects, US scientists had discovered 'no statistically significant evidence of an overall increase in annual temperature or change in rainfall in continental USA from 1895 to 1987' (Gosling, 1990, citing Newell, 1989). Similarly Karl, of the Massachusetts Institute of Technology, was reported to have found that there has been 'little or no change in world sea surface temperatures since 1850' (Gosling, 1990, citing Karl, 1988).

The picture will remain confusing until it is resolved by better data, especially that based on measurements from satellites. Precise measurements of atmospheric temperature have only been possible since launching of a new generation of satellites in 1979. Microwave radiometers attached to these satellites can measure world-wide temperatures within a day, and cover remote land areas and oceans as well as more populated areas.

Spencer and Christy (1990) have reported on the satellites programme conducted by the National Oceanic and Atmospheric Administration in the US. They show how the radiometers used enabled a precision of 0.01 degrees Celsius. Such precision is well within the plus or minus 0.1 degrees Celsius necessary for accurate climate monitoring. For the decade to 1988, during which the satellite programme had been operating, they found no obvious trend in temperatures. Over a short period of ten years, cyclical climatic phenomena, such as the 11 year cycle of solar activity, can hide longer term trends. A longer series of satellite data (at a minimum another ten years), is needed before firm conclusions can be drawn about likely greenhouse trends.

Forecast increases in global temperatures through GHGs rely on computer models which simulate physical climate relationships. The absence of consistent long-term empirical data to support model forecasts seriously undermines their validity. Nevertheless, such forecasts form the basis of most calls for international action to counteract the greenhouse effect. For example, the United Nations Environment Program's Intergovernmental Panel on Climate Change (IPCC, 1990) estimates that emissions of greenhouse gases will lead to a global warming of 0.3 degrees Celsius per decade up to the year 2100 under its 'business as usual' scenario. These estimates are based on complex General Circulation Models (GCMs) and simpler box diffusion models.

GCMs are three dimensional mathematical models which simulate physical processes in the atmosphere or the oceans. Ocean and atmospheric models can be coupled together to simulate climatic processes more fully, but require highly sophisticated computer technology to be run effectively. Box diffusion models can be used as a simpler approximation. However, the CSIRO (Source Angus McEwan (1990), Chief, Division of Oceanography, pers. comm.) has criticised the theoretical constructs of the Box Diffusion models used by the IPCC and reiterated the importance of unknown variables in the IPCC report. These unknowns include: cloud behaviour; the behaviour of the oceans; the sources of the emissions and importance of sinks; and effects on polar ice sheets.

The behaviour of these unknowns has major implications for all climate modelling. Of great significance is the effect of differing assumptions about increased cloud cover through higher levels of water vapour in the atmosphere. Water vapour is by far the most important greenhouse gas. Increases in global temperatures would lead to increased evapotranspiration, and much of the increase in global temperatures forecast by climate models results from the positive feedback of higher levels of water vapour in the atmosphere. However, once cloud formation is allowed for, the effect is not nearly as large. It could, in fact, be negative depending on whether increases in water vapour result in high or low level cloud formation. High level clouds bring a warmer atmosphere because they reflect less sunlight and emit less infra-red radiation out to space. Low level clouds have the opposite effect, cooling the atmosphere. The UK meteorological office found that introducing cloud formation into its model *reduced* estimates of the rise in global temperatures from 5.2 degrees to 1.9 degrees Celsius. Consequently, their model estimates moved from being the highest estimates of temperature increase made by large scale models to the lowest (Mason, 1989).

Scientists are by no means agreed on the size of global warming, or even its existence. Unfortunately, bureaucratic processes have translated strongly qualified estimates into what purport to be facts. In the words of the IPCC report, 'Although scientists are reluctant to give a single best estimate ... it is necessary for the presentation of climate predictions for a choice of climate predictions to be made' (IPCC, 1990: 15).

The 0.3 degree Celsius increase per decade chosen by the IPCC is not a scientific estimate but a consensus opinion decided by committee. Ellsaesser (1989) discusses how decisions made by committee 'transforms preliminary guesstimates into a consensus supported by a constituency with a vested interest in confirming and perpetuating the problem rather than solving it' (Ellsaesser 1989: 67). Scientific observation and modelling has not proved the existence of the enhanced greenhouse effect. Perhaps over-cynically, some have suggested that there is a highly articulated demand for global warming by committees and researchers but no evidence of an actual physical supply.

The effects of increased greenhouse gas emissions

It is important to emphasise the weak foundations upon which the global warming theory is postulated. Such caveats are required to avoid giving tacit support to an unproven theory. However, given the public concerns greenhouse matters have engendered, it is necessary not only to assess the theory, but also to begin asking what are the likely effects (should it be true) and what policies (if any) should be pursued?

Temperature

The immediate effect on personal comfort of a three degree increase in global temperatures over the next 100 years would not be significant. As Ellsaesser (1989) has pointed out, few farmers, businessmen, or investors would alter their behaviour after being told that the mean temperature may possibly rise by that amount over the next 50 to 100 years. Daily temperature fluctuations are well in excess of three degrees Celsius, as is the level of error in daily weather forecasts. In a life-time, many individuals will migrate between areas with at least that difference in average temperatures. Admittedly the change is an average, hiding potentially larger regional and daily variations. However, a brief consideration of the likely pattern of variation helps answer these concerns.

Most of the warming would take place during the night when greenhouse gases prevent infra-red radiation from escaping. Increasing night temperatures will increase the number of frost-free days and give longer growing seasons. In the middle latitudes, a one degree rise in average summer temperatures increases frost-free growing seasons by approximately 10 days (Kellogg, 1989).

Temperatures at the Equator would vary little, but far larger increases would be experienced nearer the poles. Colder climates could become warmer making them more comfortable to live in; but the hotter equatorial climates could be no more uncomfortable

Sea level changes

Accompanying their forecast temperature increases, the IPCC estimated increases in sea level of six centimetres per decade. Many uncertainties surround these forecasts. Beckerman (1990) points out that estimates of possible sea level rises have fallen dramatically over the 1980s. In 1980, figures as high as eight metres by the end of the next century were being cited. By 1989 one metre was the accepted wisdom. The IPCC estimate is lower still; and, the rise in actual ocean levels (less than two centimetres per decade) has only been one third of the IPCC model predictions. Beckerman also calculates that the cost of building sea walls to counteract a one metre rise is trivial when placed in a hundred year time horizon.

Sea level changes would mostly result from water in the oceans expanding with heating. Earlier fears of a collapse of the polar ice caps are now thought to be unfounded (Mason, 1989). Paradoxically, increases in precipitation could lead to *increased* ice and snow cover.

Precipitation

Should global warming be occurring, there will be also increased global precipitation. In Kellogg's (1989) scenario of changes in precipitation, countries with drier climates would be the main beneficiaries. These would include Australia, India and North

Africa. Central parts of North America may experience lower precipitation levels. In addition, Kellogg's scenario suggests that polar regions will be dryer and increases in polar ice and snow may not occur. All aspects of the scenario are, however, subject to the limitations in the modelling procedure which are discussed by Ellsaesser (1989).

CO_2 productivity

Carbon dioxide and water are the main material inputs to photosynthetic processes. Singer (1989) has likened CO_2 to 'plant food'. The IPCC acknowledges this effect stating 'Enhanced levels of carbon dioxides may increase productivity and efficiency of water use of vegetation' (IPCC 1990 : 2)

Beckerman (1990) discusses how US Water Conservation Laboratory experiments have shown large increases in plant growth with increased CO_2 and the same amount of water. Alternatively, the same amount of plant growth could be achieved with less water. He also cites studies by the Environmental Protection Authority (EPA) in the US, which estimate how the positive effects of increased carbon dioxide concentration will compensate for the negative effects of a dryer North American interior. The net effect on US agriculture could range between plus or minus $10 billion. The worst case amounts to a loss of only 0.2% of total national product.

On a global scale, Kellogg (1989) takes the view that plant growth will increase 5% on average by the year 2000. He claims the estimated gain so far this century is 10% or more.

Greenhouse policies

Should any policies be introduced at this stage?

Aside from the continuing uncertainty about the existence of the greenhouse effect, a major issue is whether it would be appropriate to attempt its arrest. It is, of course unfashionable to view man's effects on the environment as being anything other than harmful. Yet the preponderance of the changes we have discussed are likely to be beneficial—at least to ourselves. And whilst it is true that net benefit has rarely occurred where the changes have been unintentional, such outcomes would not be totally unknown. To what degree, for example, is the world's fish stock increased as a result of the expansion of ocean nutrients which man's presence has brought about? The much beloved countryside in Western Europe was largely created by man cutting down forests for agriculture and, until the eighteenth century, to manufacture charcoal.

If projections of a greenhouse effect are well founded, the greater precipitation and higher levels of carbon dioxide would, in net terms, allow higher levels of food production. Almost certainly, these benefits would be sufficient to offset the deleterious effects of climate change in some areas and the inundation of certain islands. Combating global warming would require quite substantial shifts in arranging production and, on present estimates, significant reductions in living standards. Dismissing the beneficial implications, one body of opinion calls for urgent pre-emptive action—even in the absence of conclusive evidence—since awaiting such evidence may require higher or even prohibitive costs in the future. Whilst this line of reasoning is plausible, the long list of previous forebodings which have proven unfounded tells us that, had we taken the advice of those offering it, the world today would be much the poorer for it. Some caution against precipitate and costly action is therefore warranted.

The alternative to the pre-emptive, anticipatory approach is the adaptive or reactionary approach. Here, countervailing strategies would only be implemented when, and if, the greenhouse effect becomes a problem. They could involve such action as building sea walls and providing support for relocating population, agriculture or industry. The old adage is that prevention is better than cure. But, while the risk of disease is uncertain, the diagnosis hypothetical, and the cure potentially inexpensive, adaptive cures could be preferable. To extend the analogy, the occasional band-aid and runny nose may be preferable to perpetual hospitalisation of the healthy.

An approach which incorporates elements of both pre-emptive and adaptive policies is the 'no-regrets' policy suggested by White (1990), Schneider (1989), the Australian Treasury (The Treasury

1989) and others. In the absence of conclusive evidence for a greenhouse effect, a 'no regrets' policy would first implement environmental and economic policies which are sensible in their own right. When more conclusive information on the effects of GHGs becomes available the appropriate action, be it pre-emptive or adaptive, could be taken.

The Treasury (1989) has identified four areas in the Australian economy where sound economic policies could have a beneficial impact on GHGs. The first relates to efficient management of forest resources which are net absorbers of CO_2. However, as we noted in Chapter 5, the best forest policy option for absorbing carbon dioxide is not forest preservation. Instead, trees should be harvested on a regular cycle encouraging higher growth rates and younger forests which have higher rates of carbon fixation. Such a management strategy would reduce atmospheric carbon, as long as the wood harvested from forests is used as building materials and not rotted away. Of course, in the end, wood products will decay; but holding carbon in wood products for long periods of time reduces the rate of net emissions.

The second 'no regrets' policy area is correct market based pricing and investment criteria for electricity utilities. Fossil fuel based electricity production contributes some 44% of anthropogenic CO_2 emissions in Australia (Marks, et al. 1989). The Treasury (1989) suggests that subsidised electricity generation has resulted in massive investments in large power stations through the acceptance of lower than market rates of return. Similarly, the Industries Commission (IC, 1989) has found that electricity tariff structures do not encourage efficient consumption. It is possible that corporatisation or privatisation of state electricity commissions could alter investments in power generation capacity and consumption patterns in favour of reduced CO_2 emissions. However, it could also result in lower electricity costs and prices and increased consumption. Conclusions about the net effect of market oriented energy policies on GHGs should be treated with caution.

The third 'no regrets' policy area could be in transport regulation and pricing. The transport sector is the second largest contributor to anthropogenic CO_2 emissions, being responsible for 21% of total emissions (Marks et al. 1989). The Treasury suggests that removing inefficiencies in coastal shipping and the waterfront would encourage a move from land to water based transport of freight. Removing inefficiencies in railways and a better system for charging heavy vehicle road usage may effect a shift from road to diesel based rail transport, with consequentially lower CO_2 emissions. However, this would be partly offset by the suggested removal of rail subsidies and removal of restrictions on the use of road transport.

Finally, lowering effective rates of assistance in the agricultural sector could also impact on GHG emissions. The overall rate of assistance to agriculture is 9%, which is lower than in many of Australia's trading partners. But, based on IC estimates, the Treasury identifies very high rates of assistance in two sub-sectors which generate high levels of methane emissions. Dairying has a 55% effective rate of assistance, and rice a 50% effective rate. From a global perspective, if other countries also reduced rates of protection, Australia could end up producing more of these products. However, the world cost would be raised and, with higher prices, demand would be lower. But, again, it is difficult to draw firm conclusions that such measures would have a positive impact in reducing GHG emissions because the displaced demand might even re-emerge in areas of consumption and production which result in equivalent effects.

Exploration of possible policy approaches involving costs

Though the impact of 'no regrets' policies on GHGs is not easily determined, there are other compelling reasons for pursuing them. Deciding on pre-emptive or adaptive polices over and above this response is far more difficult. However, if we assume that global warming will occur, two guidelines for directing policy decisions can be identified. First, the costs and benefits of alternative policies need to be carefully evaluated. Ideally, evaluations should be made for different levels of greenhouse gas reductions rather than being constrained to a given target. Secondly, where possible, market mechanisms should be used to enable any policy to be implemented at minimum cost. Market mechanisms allow flexibility in technology choice and create incentives to choose the most efficient means of achieving reductions in GHGs, if this is desired.

Evaluating costs against a range of negative greenhouse benefits

The costs of pre-emptive greenhouse policies are not small. Marks et al. (1989) estimate that for Australia to reduce carbon dioxide emission by 20% by 2005, in line with the Toronto target, would mean real increases of electricity tariffs by at least 41% and fuel prices by somewhere between 84% and 169%. Real wage levels, which in the base case are estimated to grow by 0.29% per annum (the low rate reflecting the need to stabilise overseas debt), would decline by 0.18% per annum. Thus an aggregate increase in real wages over the seventeen years of 5% would become, instead, a decrease of 3%. Many of the premises Marks et al. use are designed to present conservative cost figures. For example, technological improvements in fuel efficiency are assumed. Nonetheless, the total cost, expressed as the present value of GDP foregone, would be $31.7 billion in 1990 dollars at a five per cent discount rate, or 8.4% of GDP.

Some studies have contested these findings and claimed positive economic gains from implementing the 20% target (for example, see the report by Deni Greene Consulting Services, 1990). Such gains are purported to come from using less costly forms of energy. Before their claims can be credible, those attempting to justify GHG targets as a 'costless'—or indeed cost saving—measure must demonstrate compelling reasons why individuals have not already chosen such technologies. Government manipulation of peoples' behaviour forces them to make choices they would not otherwise make, and usually leads to losses in individual welfare. Apart from removing existing government distortions to market signals, it is doubtful that stringent policies to achieve greenhouse targets can be achieved without significant economic costs. Many overseas studies confirm this view (see Nordhaus, 1990, for a review).

The measures by Marks et al. focused on the cost of achieving a given level of CO_2 reduction. A more comprehensive approach would be to calculate the costs and benefits of achieving different levels of reduction. Small reductions in GHGs, can quite likely be achieved at minimal cost. Additional reductions will impose an increasing burden on society. If a target is to be imposed it should be set at the point where the costs of an extra unit of abatement just equals any net benefits obtained from GHG reductions.

Nordhaus (1990) has attempted to place current estimates of the costs and benefits of different levels of GHG reduction within a standard conceptual framework. This allows each action's response to be calibrated so that a menu of different measures can be selected from, and costs and benefits equilibrated. He estimates that the benefits from GHG reductions only justify a 10% reduction in current GHG emissions measured in carbon dioxide equivalents. The 10% reduction allows for all economic costs identified in the study and uses a discount rate 1% above the rate of growth in output. Phasing out CFCs together with a very small reduction in CO_2 emissions would be sufficient to achieve the reduction. Even if a higher level of damage is assumed (Nordhaus' 'medium' damage function), the efficient level of reductions in current CO_2 emissions would only be 6%. This is much lower than the Toronto target of 20% below 1988 levels. The implication is that the costs of major reductions in CO_2 emissions far outweigh any hypothetical benefits from reducing global warming.

The costs of an adaptive approach are potentially far smaller than pre-emptive policy targets. In addition, higher economic growth from postponing precipitate actions, provides an expanded income base out of which adaptive costs can be paid. Although we have seen no estimates of adaptive costs for Australia, some overseas calculations have been made. As previously addressed, total losses in the US agricultural sector from doing nothing at all about GHGs are only 0.2% of national product, under a worst case scenario (Beckerman, 1990). In Australia, where rainfall is liable to increase, there is more likely to be a net benefit to the agricultural sector. Based on EPA data, Beckerman estimates that the full once only capital cost of building sea walls to protect US cities from a one metre rise in sea levels by 2090 would be only 0.27% of one year's GDP. Beckerman also cites Cline (1989) as estimating that the total costs of building sea walls world wide, plus the value of land lost in Bangladesh, amounts to only 1.06% of estimated global GDP in 2090. As he points out, over a 100 year period, such costs are clearly affordable. Problems of compensating losers do but, assuming pre-emptive abatement cost are much higher, it would benefit all countries concerned to negotiate a compensation agreement.

Using market mechanisms to minimise costs

Despite the evidence that setting greenhouse targets is at best premature, many countries including Australia have begun to move in that direction. If such a pre-emptive policy is to be adopted, market mechanisms must be employed to minimise its undesirable economic impact. The ability of a market for greenhouse emissions to develop automatically seems improbable, perhaps as unlikely as in the classic case of defence. Accordingly, governments must determine acceptable levels of outputs for these by-products of production.

Ideal approaches for greenhouse gases have been previously alluded to. They would be to establish a base of globally permitted outputs and then determine the levels of allowable emissions either:

- by reference to existing usages say in each country; or
- by reference to population levels.

The former has the advantage of minimising the impact of collective decision making on the income generation process. The latter, however, starts from the basis that all people have an equal right to live.

The application of market processes would require that emission trading be permitted within nations, between nations (or preferably between firms and individuals of different nations), and between gases—the value of which would be negatively weighted in accordance with their greenhouse enhancing characteristics. If buyers are free to search out supplies, and their owners free to dispose of them at the best price, we have the ideal circumstances for efficient trade to take place.

A market in greenhouse gases, tapered towards reducing permissible levels of outputs, would limit/change the disbenefits they generate, but at a lower cost. Such a market would remove the necessity for arbitrary bans on certain sources of GHGs the price mechanism would squeeze out those offering more damage per dollar.

The command and control solution would assign permits to each supplier and each user's needs would be carefully vetted. The market approach simply allocates the total quantity of acceptable incremental carbon and leaves the price system to determine which users and uses will be accommodated. Incentives are generated both for users to economise on the substances and for suppliers to provide those which generate least relative harm.

Unfortunately it is unlikely that such a pure system will be allowed to operate. In the case of CFCs, 'political realities' have forced governments—as diverse in ideology as those of the UK and the States of Victoria and Tasmania—to insist upon certain overarching regulations. In the main these regulations seek accelerated phase-out of uses like aerosols, for which substitutes are most readily available. Though such requirements accord broadly with economic substitutability, political requirements will be much less flexible than those of the marketplace; exceptions within the targeted categories will be necessary—for example, in the case of aerosols and for certain specialised medical usages. Avoidable administrative and lobbying costs will be incurred.

The experience with Australia's Ozone Protection Act does, however, provide interesting corroboration of the power of market systems to allocate goods in a cost effective manner. The Act requires a 50% reduction in usage, with tradeable permits having been given to those supplying CFCs and halons in proportion to their 1986 production/import levels. An accelerated phase out of aerosol usage (about 30% of 1986 demand) was the main mandatory feature, and a tax of $1.06 per kilogram was imposed to cover administration costs. Aerosol usage in 1990 was down to 500 tonnes or 6%. The price of the gases rose from about $2.50 per kilogram to $6 by the third quarter of 1990, and some 40% of the original quota allocation has been traded. Although there are obligations on firms to recycle, these are not policed. Firms have, instead, taken active steps to conserve and re-use the chemicals internally; and devices are being marketed to reduce waste. Technological adaptation has been encouraged by the price rise; and

**Chart 10.3
Potential gains to co-operation**

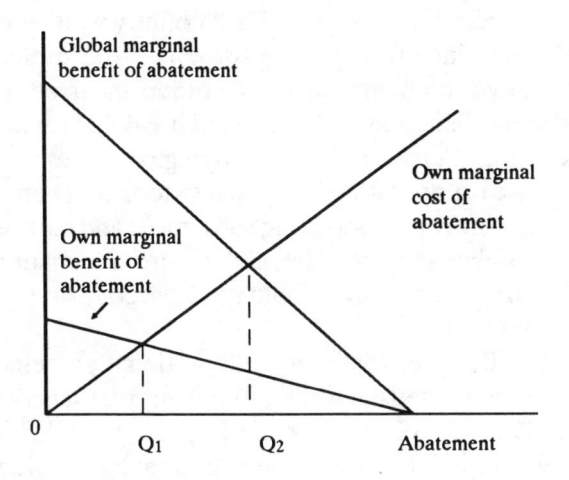

[Figure: Graph with axes showing Abatement on horizontal axis. Curves labeled "Global marginal benefit of abatement", "Own marginal cost of abatement", and "Own marginal benefit of abatement". Points Q₁ and Q₂ marked on horizontal axis.]

Dupont has, in fact, developed a new product in Australia to replace frion 12—the normal gas used in automobile air conditioning.

Prospects for inter-government agreements

A system of tradeable rights to GHGs, similar to that of CFCs, could be adopted to combat global warming should a scientific consensus emerge that such action is necessary.

This is not to deny that it would be considerably more difficult to arrange equitable transfers to prevent global warming. Even if the overall effect is proven—and proven to be negative—countries like Australia, along with Canada and most of the Soviet Union, would gain on balance. A crucial question is then: what is the likelihood of achieving global agreement on reducing GHG emissions, should it be desirable?

Devising acceptable payments for the global externalities generated presents some of the most difficult compensation arrangements imaginable. Ownership of the upper atmosphere at the present stage is impossible to envisage. Drawing upon Demsetz (1976) and Hardin (1968), the likely conclusion is that agreement to internalise global property externalities will be impossible; whilst proposals for countries to act in concert will fail as a result of free-rider problems. Each country would do better for itself if it refrained from joining (or joined and subsequently reneged on) any agreements, providing others incurred the costs of keeping them. Sanctions on agreement breakers would appear unlikely to work. Furthermore, developing countries would see much less gain in cooperating if the system took as given existing usage levels of, say ozone depleting substances and carbon emissions. Their application would be all the more trenchant as their own usage of these substances would not have contributed to any problems generated.

Barrett (1990) notes that some international agreements do seem to be operating satisfactorily without any real sanctions. He regards the 1946 Convention for the Regulation of Whaling to have been woefully ineffective; even so, whaling activities have been sharply curtailed and previously endangered genotypes of whale are increasing in number. He considers the more successful agreements to have been: the 1987 Montreal Protocol on Substances that Deplete the Ozone Layer; the World Heritage Convention; and the European 'Thirty Per Cent Club' to reduce transborder sulphur emissions. To these could be added limited but significant areas where the world community has adopted sanctions—for example, financial and other sanctions against South Africa, banning arms supplies to certain 'renegade' states, and preventing trade in narcotic substances and (probably misguidedly) in certain endangered species.

Barrett attempts to model how far countries would find it in their interest to undertake abatement activities. In Chart 10.3, each country is assumed to be identical and able to choose a level of abatement whereby its marginal costs equal its marginal benefits. If Q_1 is the optimal level of abatement for each country (chosen on the basis that all others would co-operate in a similar fashion), these choices would sum to Q_2, which would be the optimal global level of abatement.

As Barrett notes, these gains from cooperation would be frustrated by free-riding—where the countries' own marginal benefit is achieved whether or not it incurs the necessary costs itself. Indeed the outcome could be inferior to that preceding any failed agreement. The prospects of achieving cooperation are stronger where each country's costs of undertaking abatement are low and the benefits from such action are high.

Thus, drawing from US estimates, the global costs of winding down usage of CFCs are of the order of $200 billion, whilst the benefits from preserving the ozone layer are arguably many times this. The costs of dispensing with ozone depleting substances or substituting for less destructive forms are relatively low. Given all these factors, workable levels of global cooperation appear likely to be achieved; but this is much more difficult to envisage with carbon emissions. Aside from the greater uncertainty concerning the deleterious effects of burning fossil fuels (of which coal is regarded as the most injurious), it is considerably more expensive to take evasive action in shifting out of coal and lowering production generally. This is all the more likely given the less certain and highly variable inter-country consequences of greenhouse.

For the generality of unownable environmental goods, scepticism is warranted about the prospects for policies which are desirable in an aggregate sense but inimical to each country's interests when examined in isolation. If a country like Australia can only make a 1% contribution to defraying a common problem, action on which would require it to incur costs, then lasting agreement appears unlikely. Sticking to an agreement, where sanctions may not be imposed and where any benefits can be reaped without adherence, appears to make little sense if nations' overwhelming motivation are self-interest.

And yet, powerful though the pursuit of individual self-interest is, it is not people's exclusive motivating force. It is possible to observe a considerable number of actions, ranging from voting in elections through voting for political parties against the voter's self interest, to voluntary and anonymous charitable donations—all of which are difficult to place within the economist's rubric. In addition to such ostensibly altruistic actions (perhaps motivated by self satisfaction), are those where esteem is obtained from conduct seen by others as making worthy sacrifices. On an individual level, we have already seen actions with regard to CFCs where people have voluntarily chosen less efficient aerosols out of concern for the environment. Commercial firms have also voluntarily undertaken expenditures of an environmental preserving nature. This also may in part be more conventionally motivated. Mitsubishi was clearly alarmed by proposed boycotts of its product range, advocated by green activists because of the involvement of one of its subsidiaries in Indonesian logging.

Most calls by interested parties, which appeal to some greater good and appear to be made notwithstanding vested interest, contain at least an element of hypocrisy. This is clearly the case in such matters as tariffs and occupational licensing. It is equally the case where neighbours seek suppression of emissions at a factory which was in place before they arrived. In order to gain support, advocates appeal to a wider audience; such support will be forthcoming where the target group loses little from the intervention and may indeed see it as creating a precedent for analogous action which it may itself wish to obtain in the future.

Whatever the motivations, willing sacrifices are made by individuals. It is equally clear that nations too will adopt policies which are seemingly irrational given known free-rider opportunities. In some cases self interest may play a part. It could be argued that World Heritage Listing adds to a country's tourist attractions—though it is doubtful that it would do so to the extent of Australia's substantial listings. A nation might, like a private individual, forgo obvious income earning capacities in one direction for fear of retaliation on the generality of its products. But even this requires an explanation of why the previously willing buyers of those other products would forego opportunities for satisfying their needs at the least cost. South Africa has suffered from lower prices for most of its exports because potential customer nations were willing to deny themselves the most advantageous supply; financial sanctions on South Africa while causing that country to incur additional costs have only been possible because international savers and bankers have willingly accepted somewhat lower profits than they might otherwise have obtained.

In many cases, there can be no other interpretation of a nation's actions in denying itself income than that it *is* motivated by altruism, in spite of this being combined with the kudos of approval from other nations. Even if apparent altruism on the part of a particular party is mixed with a narrower form of self interest, it is nonetheless present. Clearly, narrow self interest (which is the engine driving most economists' paradigms) is not as robust as the profession sometimes imagines it to be.

The mixture of incentives other than narrow self interest which motivate people in modern societies may well be less potent than those found in tradition directed societies which preceded the modern era. These other considerations may also be less universal within the present company of nations than they were in the close-knit societies that preceded them. They are, however, clearly potent and sufficiently well observed for certain applications; and may become even more so as the world is no longer divided into Socialist and Capitalist camps.

The forgoing is not intended to deny the potential for strategic behaviour by countries attempting to free-ride on another nation's altruism. It simply indicates that where global problems are clearly manifest, agreement can be reached. The agreement by signatories to the Montreal Convention to reduce ozone depleting emissions is a clear example. It remains to be seen whether or not this may set a pattern for GHG reduction, should such measures be deemed warranted.

Some suggestion on the nature of a global treaty

Quesada (1989) advocates a global treaty which sets values on the (negative) economic value of carbon dioxide and other greenhouse gases and allows a mix of actions to counter the threat. Those actions could comprise regulation to reduce outputs, taxation of emissions and subsidies for global sinks to allow carbon build-up. In this he recognises the value of assigning prices and allowing trading with the prices passed on in the form of discounted costs of flooding and other deleterious consequences.

Tolba, the Executive Director of the UNEP, also sees an opportunity to make use of market mechanisms by having developed nations compensate poorer countries for not engaging in expanding their use of ozone depleting substances (Tolba, 1990). He estimates the worth of such measures (which he says should be additional to on-going aid disbursements) at $2 billion to $5 billion over the next 10 years. To raise these sums, he advocates a form of user fee to be paid into and disbursed by an international agency. In allocating the funds, he envisages particular priority would go to preservation of natural forest and other environmentally valued goods.

Somewhat incongruously, Tolba adds that the fees 'cannot, of course, be used as a licence to pollute'. That aside, his proposal has merit if it could be administered and fees collected without falling foul of the political manipulation and overstaffing which seems to be a feature of UN bodies.

Where particular countries have especial scope to engage in abatement strategies which benefit all countries, the opportunity exists for side payments to be made. Tropical forests are the most potent converters of carbon dioxide into oxygen and are largely located within poorer countries. If they must be preserved for the benefit of mankind as a whole, it is unconscionable for the rich nations (who became so partly because they cultivated or built upon their own virgin lands) simply to demand their preservation. Perhaps half of the world's rainforest is located in Brazil. If Brazil were to cease clearing this forest at its present rate, its future income would be reduced. As all countries consider they would benefit from retaining the Brazilian rainforest, it might be possible to arrange payment. Voluntary payment in the form of private endowments buying tropical forests has already taken place (see Chapter 5), though it is unlikely that such endowments will be adequately funded to undertake activity of this nature on the scale seen as necessary by many authorities.

Conclusion

Typically the greenhouse effect has been painted as a 'disaster' scenario. The often implicit assumption has been made that the risk of environmental damage is large but the economic costs of avoiding it are only moderate. Current scientific and economic evidence does not support this contention, and there is much uncertainty about the enhanced greenhouse effect. Because of these questions, rational policy responses are difficult if not impossible to make.

Prematurely opting for pre-emptive polices and setting stringent emissions targets will have large and certain costs for Australia, but have little on global GHG emissions. Australia only accounts for 1.2% of total world carbon dioxide emissions. Reducing GHGs requires a global response. While the benefits from GHG reductions are uncertain (and potentially small), the costs are large. Consequently, at this juncture, the achievement of global agreement will be very difficult, in contrast to ozone depleting gases where costs and benefits are more certain. In addition, limiting economic growth in a time of economic recession, mounting foreign debt and growing unemployment, will limit the ability to research and discover answers to fundamental question about global climate change.

Within another 10 years satellite data will have begun to either corroborate or refute the global warming hypothesis. Within the broad sweep of the hundred of years required for extensive global climate change, ten years is insignificant. Waiting that period of time will provide the knowledge and resources necessary for implementing countervailing polices. Delayed action, while pursuing active research into the problem, is the only sensible response.

Glossary

Age-class—stands of timber of the same age in a forest, or the timber taken from such stands.

Allowable cut—the amount of timber that can be harvested from a forest under the SUSTAINED-YIELD, EVEN-FLOW constraint.

Anthropocentric—having man and his welfare as the central focus of study.

Bituminous Coal—type of coal that burns with a smoky flame.

Capital—assets which are capable of generating income and which have themselves been produced. In more general use it can mean any asset or stock of assets—financial or physical—capable of generating income.

Carbon fixation—the conversion of atmospheric carbon in CO_2 molecules into organic carbon in plants through the process of PHOTOSYNTHESIS.

Chlorofluorocarbons—man-made chemical compounds of chloride, fluoride and carbon used in a wide variety of application including refrigerators and some aerosols.

Choke-off prices—in the context of CONTINGENT VALUATION, the price at which consumers are no longer willing to pay for more of a good or service.

Command and control policies—a general term used to encompass the range of direct regulatory environmental controls which mandate technical production processes; restrict the range of potential resource inputs or outputs; define particular management techniques and otherwise seek to impinge on particular aspects of production and consumption decisions.

Common property—land or property belonging to a community. With open access to all, such communal ownership rights generally lead to environmental degradation.

Contingent valuation—a form of market research which seeks to determine people's willingness to pay for goods which are normally unpriced.

Corporatisation— the creation of a legal framework in which state-owned commercial operations are obligated to pursue market-based objectives while state ownership and control is retained.

Covenants—legal instruments attached to titledeeds of ownership which limit an owners right to use or trade his property. For example, covenants may enforce particular building standards or require other parties to be notified before a property is sold.

Demand—the desire for a particular good or service supported by the necessary means of exchange to effect ownership.

Diminishing returns—an observation, often stated as a law, that as extra units of one factor of production are employed, with all others held constant, the output generated by each additional eventually falls.

Discount rate—a percentage rate used to reduce the value of future income streams and financial returns to their PRESENT VALUE. Discount rates generally reflect two elements: time preferences or the desire to consume now rather than later; and a real return on capital. A third element reflecting the riskiness of investment is sometimes added but generally risk is better handled with direct adjustments to the expected value of the future sum. Market rates of interest, which capture all the elements mentioned, are often used as discount rates.

Dissipation of rents—the using up ECONOMIC RENT by producers in ways other than lowering price.

Easements—rights of way or similar rights over anothers' ground or property.

Economic rent—The difference between the return made by a factor of production and the return necessary to bring it into production.

Efficiency (economic)—A state of the economy in which no one can be made better off without making someone worse off. There are three types of efficiency: *productive efficiency* in which output of the economy is being produced at least cost; *allocative efficiency* in which resources are applied to producing the goods and services consumers value most; and *distributional efficiency* in which output is distributed in such a way as community welfare is maximised.

Effluent taxes—taxes on liquid waste pollutants discharged into sewers and drains, or natural water bodies.

Electromagnetic spectrum the range of wavelengths of electromagnetic radiation including ultra-violet light, visible light, infra-red radiation, radio waves, gamma waves etc.

Equity—commonly associated with fairness and justice but in economics has become associated with the slightly different concept of equalising incomes or opportunities. Intergenerational equity extends this concept to equalising either income, consumption or, access to a similar stock of resources between generations.

Exclusive rights—property rights which assign full ownership to a single entity without being attenuated by legal instruments such as COVENANTS or EASEMENTS.

Expected value—a measure of the value of accruing to an investor from an asset which yields an uncertain flow of benefits. The measure is calculated using standard laws of probability.

Glossary

External costs—see EXTERNALITIES.

Externalities—The costs and benefits of a transaction between two or more parties which impact on other parties not directly involved. Sometimes called *spillover* or *third-party* effects.

Formal title—defined right of property ownership with or without possession but evidenced by legal instruments such as a title-deed.

Free-rider—an individual or group who obtains benefits without the need to make sacrifice for them.

Greenhouse effect—the raising of global temperatures by a blanket of gases around the earth reflecting infra-red radiation back to the surface.

Greenhouse gases—gases which re-radiate infra-red radiation from the earths surface creating the GREENHOUSE EFFECT. They include such gases as water vapour, carbon dioxide, methane, nitrous oxides and CHLOROFLUOROCARBONS.

Hardwood—Wood from trees classified botanically as *Angiosperms*. Most hardwood trees are broad-leaved and the wood is pored. The term does not denote the hardness of the wood, though it is sometimes used in this sense. (see also softwoods)

Income redistribution—expropriation of income by means of taxes, charges and other fiscal instruments in order further EQUITY goals through social welfare payments and the like.

Internal costs—costs which are borne by the parties involved in a transaction.

Labour/leisure substitution—the shift of activities from earning taxable income to leisure resulting from a tax on labour income. More generally it can be seen as a substitution of taxed effort by untaxed leisure or effort.

Marginal cost and benefit—the increase in total costs or benefits to a firm or organisation caused by increasing output by one extra unit.

Market-based instruments—regulatory instruments, such as taxes, charges and permits, which utilise market based incentives to achieve desired outcomes at least cost.

Mineral lease—an agreement by the owner of a mineral reserve (generally the crown) to grant rights to another party for a specified period under defined conditions.

Monopoly—a market in which there is only one supplier.

Non-excludable goods or services—goods and services are non-excludable when supply to one individual does not or cannot exclude others from also benefiting. Defence is one classic case; a scenic view is another.

Non-point externalities—externalities which are diffuse and affect a wide-range of unrelated individuals; consequently they are hard to quantify and correct.

Non-rival—used to describe a good or service which can be 'consumed' by many individuals without diminishing its value. (see also NON-EXCLUDABLE—a related but slightly different concept)

Opportunity costs—The value of alternative uses or activities which must be given up to acquire or achieve something else of value.

Optimum—a position in which the aim of any economic unit is being served as effectively as possible within any constraints applying. It corresponds to economic *efficiency*.

Photosynthesis—the process in which the energy in sunlight is used by green plants to build complex organic substances from carbon and water.

Point externalities—externalities which affect only a few parties and consequently are amenable to negotiation and common-law agreements.

Present value—the value of a future financial sum or stream of returns discounted to its value in current dollars at a chosen DISCOUNT RATE. If the chosen discount rate was the market interest rates, then the present value would be the current sum one should deposit in the bank now to yield an equivalent future sum.

Private costs—costs borne by the individuals involved in a production or consumption decision. (compare SOCIAL COSTS)

Privatisation—principally, the sale of government-owned commercial enterprises to private investors, with or without loss of government control in these organisations.

Property rights—rights of ownership. For economic EFFICIENCY they need to be private, defined, monitored and tradeable.

Pulp-log—logs unsuitable for sawmilling but used in the production of wood chips, pulp and paper and wood panels.

Resource rent—see ECONOMIC RENT

Resource rent tax—a tax supposed levied on the economic rent associated with a resource such as a mineral deposit.

Ricardian rent—see ECONOMIC RENT

Riparian rights—rights to water assigned on the basis of who owns the river-bank.

Risk class—a group of investments which involve a similar level of risk.

Risk premium—an adjustment made to a DISCOUNT or interest rate to allow for uncertainty.

Market, Resources and the Environment

Royalty—a sum paid to a resource owner for use of the resource.

Salination—Increased levels of mineral salts high in the soil profile caused by rising water tables. There are two types of salination: *dry-land salination* caused by clearing of trees and vegetation with increased water run-off raising water tables downstream; and *wet-land salination* caused by irrigation.

Saw-log—logs suitable for sawmilling.

Social costs—costs of a production or consumption decision borne by individuals and communities not directly involved in such decisions. (compare PRIVATE COSTS)

Softwood— Timber from conifer species such as radiata pine and cypress pine.

Spillover costs —see EXTERNALITIES

Supply —the quantity of a good (or) service available for sale at any specified price.

Sustainable development—a broad term popularised by the Brundtland Report, *Our Common Future,* in 1987. They defined it as 'development that meets the needs of the present without compromising the ability of future generations to meet their own needs'. It is thus closely related to the economic concept of intergenerational EQUITY.

Sustained-yield, even-flow management—a traditional forestry management policy which maximises an even-flow of timber volume over time.

Tradeable rights—see TRADEABLE QUOTAS AND PERMITS

Transaction costs—costs involved in economic activities such as the sale and purchase of goods which do not in themselves contribute to the value of the activity. For example: negotiation costs, legal costs, etc.

Transferable quotas and permits—rights to emit pollutants which can be exchanged between firms, and individuals. A global ceiling is generally set on the total number of permits or quotas issued.

Vertical aggregation of the demand curve—a summation of the prices all individuals would be willing to pay for each unit of a good or service.

Vest—confer formally on an individual or group of individuals an immediate fixed right of present or future possession.

Welfare losses—losses in consumer and producer welfare which generally result from the imposition of a government regulation or tax which forces consumers and producers to arrange their affairs in ways they would not freely choose. As a simplistic illustration, if an individual is *forced* to eat an orange instead of an apple, his welfare is diminished even if both cost the same.

Wildcatter—speculative oil explorer.

Work program bidding—offering undertake expenditures as a condition of obtaining a lease.

Acknowledgments to: Bannock, G., Baxter, R.E. and Davis, E. (1987)*Dictionary of Economics* 4th ed. London: Penguin Books; and *The Concise Oxford Dictionary* (—6) 6th ed. Oxford: Oxford University Press

Abbreviations

R&D	Research and Development
IAC	Industries Assessment Commission — now the IC (Industries Commission)
ACF	Australian Conservation Foundation
OECD	Organisation for Economic Co-operation and Development
SEFA	South-East Forest Alliance
NAFI	National Association of Forest Industries
CSIRO	Commonwealth Scientific and Industrial Research Organisation
ACE	Allowable cut effect
CFCs	Chloroflorocarbons
GHGs	Greenhouse gases
APPM	Australian Pulp and Paper Manufacturers
AFH	Associated Forest Holdings—a division of APPM
GDP	Gross Domestic Product
GNP	Gross National Product
ADR	Australian Design Rule
EPA	Environmental Protection Agency

References

Chapter 1

Grilli, E.R. and Yang, M.C. (1987) 'Primary Commodity Prices, Manufacured Goods Prices, and the Terms of Trade of Developing Countries: What the Long Run Shows', *The World Bank Economic Review*, Vol.2, No. 1.

Industry Commission, (1990) *Measuring the Performance of Selected Government Business Enterprises*, Canberra, August.

Chapter 2

Bernstam, M.S. (1989) 'Productivity of Resources, Economic Systems Population and the Environment:' *Is the Invisible Hand Too Short or Crippled?*, Centre of Policy Studies, Monash. To be included in Davis, K. and Bernstam M.S. (1990) (eds.) 'The Endless Frontier and Resources', Cambridge Univ. Press, New York.

Boyet, W.E. and Tolley, G.S. (1966) 'Recreational Projections Based on Demand Analysis', *Journal of Farm Economics*, 48(4).

Coase, R. (1960) 'The problem of social cost', *Journal of Law and Economics*, 3: 1–44.

Krutilla, J.V. and Fisher, A.C. (1985) *The Economics of Natual Environments: Studies in the Valuation of Commodity and Amenity Resources*, Washington DC: Resources For the Future.

Pearce, D., Markandya, A., Barbier, E.B. (1989) *Blueprint for a Green Economy*, London: Earthscan Pulications.

Pearce, D. (1980) 'The Social Incidence of Environmental Costs and Benefits', *Progress in Environmental Planning and Resource Management*, 2.

Sowell, T. (1987) *A Conflict of Visions*, New York: William Morrow and Company Inc.

Wildavsky, A., 'No Risk is the Highest Risk of All', in Clickman, T.S. and Bough, M., *Readings in Risk*, Washington DC: Resources For the Future, 120–128.

Chapter 3

Andrews, John (1966) *Frontiers and Men: A volume in memory of Griffith Taylor*, Melbourne, F.W.Cheshire.

Bach, John (1982) *A Maritime History of Australia*, Sydney, Pan Books.

Barrett, Bernard (1971) *The Inner Suburbs: The evolution of an industrial area*, Melbourne, Melbourne University Press.

Bird, Eric (1985) 'The future of the beaches', in Heathcote (1985)

Blainey, Geoffrey (1963) *The Rush that Never Ended: A History of Australian Mining*, Melbourne, Melbourne University Press.

Blainey, Geoffrey (1966) *The Tyranny of Distance*, Melbourne, Sun Books.

Blainey, Geoffrey (1975) *Triumph of the Nomads: A History of Australia*, Melbourne, Macmillan.

Cain, N. (1962) 'Companies and squatting in the Western Division of New South Wales', 1896–1905, in Barnard (1962).

Campbell, K.O. (1970) 'Land Policy', in Williams (1970) Chap 8.

Campbell, Keith O. (1980) *Australian Agriculture: Reconciling change and tradition*. Melbourne, Longman Cheshire.

Carron, L.T. (1985) *A History of Forestry in Australia* Canberra, Australian National University Press.

Davidson, B.R. (1969) *Australia Wet or Dry?: The physical and economic limits to the expansion of irrigation*, Melbourne, Melbourne University Press.

Dingle, Tony (1984) *The Victorians: Settling*, Sydney, Fairfax, Syme and Weldon.

Duncan, Ross (1967) *The Northern Territory Pastoral Industry 1863–1910*, Melbourne, Melbourne University Press.

Doran, C. R. (1984) An Historical Perspective on Mining and Economic Change, in L.H. Cook and M.G. Porter (eds.) *The Mineral Sector and the Australian Economy*, Allen and Unwin, Sydney

Dunstan, David (1984) *Governing the Metropolis: Melbourne 1850–1891*, Melbourne, Melbourne University Press.

Fennessy, B.V. (1962) 'Competitors with Sheep: Mammal and Bird Pests of the Sheep Industry', in Barnard (1962)

Fogarty, J.P. (1967) *George Chaffey*, Melbourne, Oxford University Press.

Heathcote, R.L. (1965) *Back of Bourke: A study of land appraisal and settlement in semi-arid Australia*,Melbourne, Melbourne University Press.

Heathcote, R.L. (1982) *Australia*, Harlow, Longmans.

Heathcote, R.L. (ed.) (1988) *The Australian Experience*, Melbourne, Longmans Cheshire.

Houghton, Norm (1975) *Sawdust and Steam: A history of the railways and tramways of the eastern Otway Ranges*, Melbourne, Light Railways Research Society of Australia.

Kearns, R.H.B. (1982) *Broken Hill: A pictorial History*, Investigator Press.

Maddox, R.& McLean I.W. (ed.)(1987) *The Australian Economy in the Long Run*, Cambridge, Cambridge University Press.

Martin, C.S. (1955) *Irrigation and Closer Settlement in the Shepparton District 1836–1906* Melbourne, Melbourne University Press.

Meinig, D.W. *)n the Margins of the Good Earth: The South Australian Wheat Frontier 1869–84*, London, John Murray.

Moore, R.M. (1962) Effects of the sheep industry on Australian vegetation, in Barnard (1962) Chap 13.

Perry, T.M. (1963) *Australia's First Frontier: The spread of settlement in N.S.W. 1788–1829*, Melbourne, Melbourne University Press.

Perry, T.M. (1966) 'Climate and settlement in Australia 1700–1930', in Andrews (1966).

Pike, Douglas (1967) *Paradise of Dissent: South Australia 1829–1857*, Melbourne, Melbourne University Press.

Powell, J.M. (1975) Conservation and Resource Management in Australia 1788–1860, in Powell and Williams (1975).

Powell, J.M. and Williams, M. (eds.) (1975) *Australian Space, Australian Time: Geographical Perspectives*, Melbourne, Oxford University Press.

Powell, J.M. (1988) 'Patrimony of the people', in Heathcote (1988),

Powell, J.M. (1988) *An Historical Geography of Modern Australia: The restive fringe,* Cambridge, Cambridge University Press.

Powell, J.M. (1989) *Watering the Garden State: Water, land and community in Victoria 1834–1988*, Sydney, Allen and Unwin.

Price, Grenfell (1966) 'The moving frontiers and changing landscapes of flora and fauna in Australia', in Andrews (1966).

Roberts, Stephen (1964) *The Squatting Age in Australia 1835-1847*, Melbourne, Melbourne University Press.

Roberts, Stephen (1968) *History of Australian Land Settlement: 1788–*

1920, Melbourne, Macmillan.

Rolls, Eric (1977) *They All Ran Wild*, Sydney, Angus and Robertson.

Sawyer, G. (1962) 'Rabbits, the Law, and the Constitution', in Barnard, (1962).

Serle, Geoffrey (1963) *The Golden Age: A history of the Colony of Victoria 1851–1861*, Melbourne, Melbourne University Press.

Smith, Adam (1958) *The Wealth of Nations*, Vol. 2, London, Dent. (Originally published in 1776).

Williams, D.B. (ed.) (1976) *Agriculture in the Australian Economy*, Sydney University Press.

Williams, M (1975) 'More and smaller is better: Australian rural settlement 1788–1914', in Powell and Williams (1975).

Williams, M. (1988) 'The clearing of the woods', in Heathcote, (1988).

Williams, O.B. (1962) 'The Riverina and its pastoral industry', 1860–1869, in Barnard (1962).

Wace, Nigel (1985) 'Naturalised Plants in the Australian Landscape', in Heathcote, (1988).

Chapter 4

ABARE (1989), *Commodity Statistical Bulletin*, AGPS, Canberra.

Balderstone, J.W. et al. (1982) *Agricultural Policy: Issues and Options for the 1990s*, Working Group Report to the Minister for Primary Industry, Canberra, AGPS.

Brett, D. (1990) 'Restoring the Murray-Darling Basin', *Ecos*, 64, 4–10.

Business Council of Australia (1990) 'Achieving Sustainable Development', *Business Council Bulletin*, August, 6–32.

Chisholm, A. and R. Dumsday (eds.) (1987) *Land Degradation: Problems and Policies*, Cambridge, Cambridge University Press.

Department of Environment, Housing and Community Development (1978) 'A Basis for Soil Conservation Policy in Australia' Commonwealth and State Government Collaborative Soil Conservation Study, 1975–77, *Report 1*, AGPS, Canberra.

Dixon, J.A. et al (1989) *The Economics of Dryland Management*, London, Earthscan.

Drought Policy Review Task Force (1990) 'National Drought Policy', Vol. 1, AGPS, Canberra.

References

Dumsday, R.G. (1983) 'Agricultural Resource Management', *Australian Journal of Agricultural Economics*, 27, 157–163.

Dumsday, R., G. Edwards and A. Chisholm (1990) 'Resource Management', in D.B. Williams (ed.) *Agriculture in the Australian Economy*, Sydney, Oxford University Press.

Eckersley, R. (1989) 'Regreening Australia: The Environmental, Economic and Social Benefits of Reforestation', CSIRO Occasional Paper No. 3.

Freebairn, J.W. (1983) 'Drought Assistance Policy', *Australian Journal of Agricultural Economics*, 27, 185–199.

Greig, P.I. and P.G. Devonshire (1981) 'Tree Removals and Saline Seepage in Victorian Catchments: Some Hydrological and Economic Results', *Australian Journal of Agricultural Economics*, 25, 134–148.

Hodge, I. (1982) 'Right to Cleared Land and the Control of Dryland-Seepage Salinity', *Australian Journal of Agricultural Economics*, 26, 185–201.

IAC (1987) *Assistance to Agricultural and Manufacturing Industries*, AGPS, Canberra.

King, D.A. and J.A. Sinden (1986) 'Influence of Land Condition and Soil Conservation on Land Values in the Manilla Shire of NSW', paper to Australian Agricultural Economics Society Conference, Canberra.

Kirby, M.G. and M.J. Blyth (1987) 'Economic Aspects of Land Degradation in Australia', *Australian Journal of Agricultural Economics*, 31, 154–174.

Milton, P. et al. (1989) 'The Effective of Land Degradation Policies and Programs', Report of the House of Representatives Standing Committee on Environment, Recreation and the Arts, AGPS, Canberra.

Molnar, I. (1955) 'Effects of Soil Erosion on Land Values and Production', *Journal of the Australian Institute of Agricultural Science*, 21, 163–166.

OECD (1975) *The Polluter Pays Principle: Definition, Analysis, Interpretation*, OECD, Paris.

Pearce, D., E. Barbier and A. Markandya (1990) *Sustainable Development: Economics and the Environment in the Third World*, Edward Elgar, Hants.

Quiggin, J. (1986) 'Common Property, Private Property and Regulation: The Case of Dryland Salinity', *Journal of Agricultural Economics*, 30, 103–117.

Quiggin, J. (1988) 'Murray River Salinity—An Illustrative Model',

American Journal of Agricultural Economics, 70, 635–645.

Robinson, I.B. (1982) 'Drought Relief Scheme for the Pastoral Zone', *Australian Rangeland Journal*, 4, 67–77.

Simmons, P. and N. Hall (1990) 'An Economic Perspective on the Allocation and Quality of Irrigation Water in the Murray–Darling Basin', Australian Bureau of Agriculture and Resource Economics, working paper.

Taylor, M., R. Dumsday and P. Bruyn (1982) *Salinity in Victoria*, papers presented at a joint Australian Institute of Agricultural Science and Australian Agricultural Economics Society Conference, Melbourne.

The Age (1990) 'Salt: The White Plague', 21 July.

Wills, I. (1987) 'Resource Degradation on Agricultural Land: Information Problems, Market Failures and Government Intervention', *Australian Journal of Agricultural Economics*, 31, 45–55.

Woods, L.E. (1984) *Land Degradation in Australia*, 2nd Ed., AGPS, Canberra.

Chapter 5

Abbott, Gordon, Jr. (1982) B. Rusmore, A. Swaney and Spader. A. D. Covello, 'Long-Term Management: Problems and Opportunities', in *Private Options: Tools and Concepts for Land Conservation*, CA, Island Press.

Anderson, T., Chisholm, A., Hartley, P. and Porter, M. (1990) 'Forests and Markets: Submission to the Resources Assessment Commission – Inquiry into Options for the Use of Australia's Forest and Timber Resources', *Occasional Paper No. B1*, Melbourne, Tasman Economic Research.

Beckerman, W. (1990) *Global Warming: An Economic Perspective; or Global Warming: A Sceptical Economic Assessment* Oxford: Balliol College, provisional draft of September.

Blood, T. and Baden, J. (1984) 'Wildlife Habitat and Economic Institutions', *Western Wildlife* No. 1 Missoula MT: Montana Forest and Conservation Experiment Station, University of Montana (Spring).

Bowes, Michael D., and John V. Krutilla. (1985) 'Multiple Use Management of Public Forestlands', in Kneese A. V. and Sweeney J. L. (eds.) *Handbook of Natural Resource and Energy Economics*, vol II. Amsterdam: North-Holland, 531–569.

Bruce, Ian A. (1986) 'Should Pines be Privatised', *Economic Papers*. 5

(1), 60–73.

Cameron, J.I., and I.W. Penna. (1988) *The Wood and the Trees*. Hawthorn: Australian Conservation Foundation.

Dodge, Sue E. (1987) *The Nature Conservancy Magazine*, 37 (5), November/December,

Dowdle, B. 1981. 'An Institutional Dinosaur with an Ace: or, How to Piddle Away Public Timber Wealth and Foul the Environment in the Process', in Baden, J. and Stroup, R. L. (eds.) *Bureaucracy vs. Environment – the Environmental Costs of Bureaucratic Governance* Ann Arbor, MI: The University of Michigan Press.

Ferguson, I.S. (1987) Dargwel, J. and Sheldon, G. Problems and Prospects in Victoria in *Prospects for Australian Hardwood Forests*, Canberra: Centre for Resource and Environmental Studies.

Flannery, T. (1989) Australian Wilderness: An Impossible Dream *Australian Natural History* 23 (2) Spring.

Florence, R. (1989) *Wood Production and the Environment: Working in Harmony* Canberra: NAFI, (National Association of Forest Industries, Ltd.).

Hardin, G. (1968) 'The Tragedy of the Commons', *Science* 162, 1243–1248.

Hartman R. (1976) 'The Harvesting Decision When a Standing Forest has Value', *Economic Inquiry*, 14, 52–58.

Higgins, Ean. (1990) Chips off the Old Block: or how to have your forest and reap it *The Weekend Australian*, 3–4 March, 29.

Hyde, W.F. (1981) 'Compounding Clear-cuts: The Social Failures of Public Timber Management in the Rockies', in *Bureaucracy vs. Environment – the Environmental Costs of Bureaucratic Governance* Edited by J. Baden and R. L. Stroup. Ann Arbor, MI: The University of Michigan Press.

(IC) Industries Commission (1990) 'Recycling in Australia – an Analysis of the Incentives to Recycle and The Prospects for Further Recycling', in *The Draft Report on Recycling* Vol. 1 Canberra.

MacArthur, Stewart. (1990) 'Barren backyard now a rainforest', *The Australian*, 14 March, 3.

(NAFI) National Association of Forest Industries (1989) *Overcoming Land Degradation : The Role of Forestry*, Canberra.

(NAFI) National Association of Forest Industries (1990) *Submission to the Resources Assessment Commission Inquiry into the Options for the use of Australia's Forest and Timber Resources* Canberra.

National Geographic (1988) 'Quietly Conserving Nature', National Geographic, December, 818–845.

Samuelson, P.A. (1954) 'The Pure Theory of Public Expenditure', *Review of Economics and Statistics*, 36 (4)

Samuelson, P.A. (1976) 'Economics of Forestry in an Evolving Society', *Economic Inquiry*, 14, 466–492.

The Council on Environmental Quality (1984) *15th Annual Report* Washington D.C.: US Government Printing Office.

The National Audubon Society (1986) *Audubon Wildlife Report*.

The Wilderness Society (1987) *Management Directions for the National Forests of the Greater Yellowstone Ecosystem* Washington D.C., The Wilderness Society.

Turner, B.J. (1987) 'Advanced Techniques for Forest Management in *Prospects for Australian Hardwood Forests',* Dargwel J. and Sheldon G. Canberra, Centre for Resource and Environmental Studies.

Walker, B.B. (1984) Management of Non-Wood Values in State Forests: Should the User Pay – 4. Funding Non Wood Values in Tasmanian State Forests, *Australian Forester*, 47 (3), 164–171.

Chapter 6

ABC 'Uncertainty Principle' 15 April 1990.

AMEC, (1990) Submission to the Inquiry into Mining and Minerals Processing by the Association of Mining and Exploration Companies, Canberra, Industry Commission.

AMIC, (1990) Submission to the Inquiry into Mining and Minerals Processing by the Australian Mining Industry Council, Canberra, Industry Commission.

Anderson T.L. and Hill P.J. 'The Race for Property Rights', The Journal of Law and Economics, Vol. XXX111 (April 1990), P177–197.

Brown E.C., 'Business-Income Taxation and Investment Incentives' in *Income, Employment and Public Policy: Essays in Honor of Alvin H. Hansen*, New York, Norton 1948

Brundtland Report (1987) 'Our Common Future', World Commission of Environment and Development, Oxford: Oxford University Press.

Church A.M. (1985) 'Natural Resource and Taxation Policies', Research Monograph no. 14, Canberra, Centre for Research on Federal Financial Relations.

Fickett A.P. and Gelling C. W. (1990) 'Eficient use of electricity', *The*

Scientific America, September: 29–36.

Garnaut R. and Clunies-Ross, A. (1975) 'Taxing of Natural Resource Projects', *Economic Journal*, Vol. 85, 284–295.

Hayek, F. (1976) 'Law Legislation and Liberty' Volume III 'The Political Order of a Free People', London, Routledge and Kegan Paul.

Hotelling, H. (1931) 'The Economics of Exhaustible Resources' *Journal of Political Economy*, April.

Industries Assistance Commission, (1976) *Report on Crude Oil Pricing*, Canberra, AGPS.

Pearce, D., Markandya, A., Barbier, E.B. (1989) Blueprint for a Green Economy, London, Earthscan Publications.

Smith, B. (ed.) (1979) *Taxation of the mining Industry*, Canberra, Centre for Resource and Environmental Studies.

Solow R.M. 'The Economics of Resources or the Resources of Economics' in Lloyd P.J. (ed.) *Mineral Economics in Australia* Sydney, George Allen and Unwin, p53–70.

Stollery K. R. 'The Discount Rate and Resource Extraction,, Resources Policy, March 1990 p47–55.

Swan P.L. (1976) 'Income Raxes, Profit Taxes and Neutrality of Optimising Decisions', *Economic Record*, 52 June: 166–181.

Vries H.J.M. de (1989) 'Effects of Resource Assessments on Optimal Depletion Estimates', *Resource Policy*, September: 253–268.

Chapter 7

Australian Broadcasting Commission (1989) 'Derwent Pollution', *Four Corners*, March 27.

Bates, G.M. (1983) *Environmental Law in Australia*, Butterworths, North Ryde, NSW.

Baumol, W.J. and Oates, W. (1988) *The Theory of Environmental Policy*, Second edition, Englewood Cliffs, N.J., Prentice-Hall.

Beder, S. (1990) 'Sun, surf and sewage', *New Scientist*, July 14.

Bolton, R.L. and Klein, L. (1971) *Sewage Treatment: Basic Principles and Trends*, Second edition, London, Butterworths.

Butlin, N.G. (ed.) (1976) *Sydney's Environmental Amenity, 1970–1975*, Canberra, Australian National University Press.

Coase, R. (1960) 'The problem of social cost', *Journal of Law and Economics* ,3: 1–44.

Deschamps, J.D. (1986) 'Privatization of water systems in France', *Journal of the American Water Works Association*, 78: 34–40.

Hahn, R. W. (1989) 'Economic prescriptions for environmental problems: how the patient followed the doctor's orders', *Journal of Economic Perspectives*, 3: 95–114.

The Sunday Herald, 'EPA to crack down on paper mill', Melbourne, September 24.

Holmes, N. (1987) 'A preliminary ecological assessment of some Victorian coastal discharges', *Environment Protection Authority Scientific Research Series*, 87/008, Melbourne, November.

Industry Commission, (1990) *Measuring the Performance of Selected Government Business Enterprises*, Canberra, August.

Kinnersley, D. (1988) *Troubled Water: Rivers, Politics and Pollution*, London, Hilary Shipman.

Melbourne and Metropolitan Board of Works, (1990) *Summary of Trade Waste Requirements*, Trade Waste Services Branch, Melbourne.

OECD (1989) *Economic Instruments for Environmental Protection*, Paris.

Russell, C.S., Harrington, W., and Vaughan, W.J. (1986) Enforcing Pollution Control Laws, Washington, D.C., Resources for the Future.

Smith, V.L. (1988) 'Electric Power Deregulation: Background and Prospects', *Contemporary Policy Issues*, 6: 14–24.

Stump, M.M. (1986) 'Private operation of US water utilities', *Journal of American Water Works Association*, 78: 49–51.

Tietenberg, T. (1988) *Environmental and Natural Resource Economics*, Second edition, Glenview, Illinois, Scott, Foresman and Company.

Chapter 8

Anderson, T.J. and Hill, P.J. (1983) 'Privatizing the Commons: An Improvement?' *Southern Economic Journal*, Vol. 50: 438–450.

Cheung, S.N.S. (1970) 'The Structure of a Contract and the Theory of a Non-exclusive Resource,' *Journal of Law and Economics*, Vol. 13: 49–70.

Clark, I.N., Major, P.J. and Mollett, N. (1988) 'Development and Implementation of New Zealand's ITQ Management System,' Marine Resource Economics, Vol. 5: 325–349.

Clark, I.N. and Duncan, A.J. (1986) 'New Zealand's Fisheries Management Policies Past, Present and Future: The Implication of an ITQ-Based Management System,' in Mollett, N. (ed.) Fishery Access

Control Programs Worldwide, Proceedings of the Workshop on Management Options for the North Pacific Longline Fisheries, Alaska Sea Grant *Report No. 86–4*, University of Alaska.

Fitzgerald, P.J. (1966) *Salmond on Jurisprudence*, London, Sweet and Maxwell.

Gordon, H.S. (1954) 'The Economic Theory of a Common Property Resource: The Fishery,' *Journal of Political Economy* Vol. 62: 124–142.

Ministry of Agriculture and Fisheries, (1984) *Inshore Finfish Fisheries: Proposed Policy for Future Management*, Wellington, MAF Fisheries Management.

Moore, O.M. (1982) 'Paua – The Impact of Amateur Catches and Illegal Harvesting,' in Akroyd, J.M., Murray, T.E. and Taylor, J.L. (comps.) Proceedings of the Paua Fishery Workshop, *Fisheries Research Division Occasional Publication No.14*, Wellington.

Murray, T. and Akroyd, J.M. (1984) 'The New Zealand Paua Fishery: An Update of Biological Considerations to be Reconciled with Management Goals', Fisheries Research Centre *Internal Report No. 5*, Wellington.

Pufendorf, S. (1688) De Jure Naturae Et Gentium Libri Octo, in Scott, J.B. (ed.) (1934) 'The Classics of International Law', Vol. 17, Oxford, Clarendon Press.

Report of the Fishing Industry Committee 1970–72 (1972) Appendix to the Journals of the House of Representatives I.14, Wellington.

Report of the Paua Shell Review Committee (1988) Department of Trade and Industry, Wellington.

Scott, A. (1988) 'Development of Property in the Fishery,' *Marine Resource Economics*, Vol 5: 289–311.

Swarztrauber, S.A. (1972) *The Three Mile Limit of Territorial Seas*, Annapolis, Naval Institute Press.

Taylor, Baines and Associates and Lincoln International (1989) Review of the Chatham Islands Economy: *Final Report*, Wellington, Department of Internal Affairs.

Chapter 9

Bernstam, M.S. (1989) 'Productivity of Resources, Economic Systems Population and the Environment:' *Is the Invisible Hand Too Short or Crippled?*, Centre of Policy Studies, Monash. To be included in Davis, K. and Bernstam M.S. (1990) (eds.). 'The Endless Frontier and

Resources', Cambridge Univ. Press, New York.

Buchanan, J.M. (1988) 'Market Failure and Political Failure', Cato Journal, 8,1: 1–14.

Coase, R.H. (1960) 'The Problem of Social Cost', *Journal of Law and Economics,* 3: 1–44.

Crandell, R.W. (1983) 'Controlling Industrial Pollution', Brookings, Washington.

Freeman, M., Haveman, R.H. and Kneese, A.V. (1973) 'The Economics of Environmental Policy', Wiley, John, N.Y.

Greiner, N.F. (1990) 'The New Environmentalism'.

Grenning, M. (1985) 'Australian Motor Vehicle Emission Policy A Costly Mistake', *CEDA Monography,* 80, Melbourne.

Hahn, R.W. and Hester, G.L. (1987) 'The Market for Bads' *Regulation,* 3,4, pp48–53.

Hahn, R.W. and Hester, G.N. (1989) 'Marketable Permits' Ecology Law Review, 16,2,pp361–406.

Hamrin, R. (1981) 'Environmental Quality and Economic Growth', Council of State Planning Agencies, Washington.

Hartley, P.R. and Porter M.G. (1990) *A Green Thumb for the Invisible Hand,* Tasman Institute, Melbourne.

Hazilla, M. and Kopp, R.J. (1989) *The Societal Cost of Environmental Quality Regulations: A General Equilibrium Analysis,* Resources for the Future.

Huber, P. (1985) 'The I Ching of Acid Rain', *Regulation,* September–November.

Levin, M.H. (1985) 'Building a Better Bubble at EPA' *Regulation,* March–April pp33–42.

Pole, N. (1973) 'An Interview with Paul Ehrlich' *The Ecologist,* 3,1: 18–24.

Samuelson, P. (1954) 'The Pure Theory of Public Expenditure' *Review of Economics and Statistics,* 36: 387–9.

Smith, F. (1989) *Environmental Policy: A Free Market Proposal,* p.32–37 Tulanian.

Terkla, D. (1984) 'The Efficiency Value of Effluent Tax Revenues', *Journal of Environmental Economics and Management,* 2, pp107–123.

Tietenberg, T.H. (1990) 'Economic Instruments for Environmental Regulation', *Oxford Review of Economic Policy,* 6,1, pp17–33.

Tolley, G.S. and Randall, A., *Establishing and Valuing the Effects of Improved Visibility in the Eastern United States,* Report to the US EPA.

Verall, F.N. and Simpson R.W. (1988) 'Trends in Ambient Air Quality in Brisbane' paper presented to ANZAAS.

Chapter 10

Barrett S. (1990) The Problem of Global Environmental Protection, *Oxford Review of Economic Policy* 6(1) Spring: 68–79.

Beckerman, W. (1990) *Global Warming: An Economic Perspective; or Global Warming: A Sceptical Economic Assessment* Oxford: Balliol College, provisional draft of September.

Beckman, P. (1989) 'What Warming', *Access to Energy* 17(3).

Cline, W. (1989) *Political Economy of the Greenhouse Effect*, Washington D.C., Institute for International Economics, preliminary draft of August.

Demsetz H. (1976) 'Toward a Theory of Property Rights', *American Economic Review* 57: 347–359.

Deni Greene Consulting Services, (1990) 'A Greenhouse Energy Strategy: Sustainable Energy Development for Australia', *A Report Prepared for the Department of Arts, Sport, the Environment, Tourism and Territories* Canberra (February)..

Ellsaesser, H.W. (1989) Response to Kelloggs Paper in S. Fred Singer (ed.) *Global Climate Change: Human and Natural Influences* New York: ICUS, pp. 67–89

Gosling, T. (1990) 'Are We Getting Warmer?', *The Herald*, 20th Feb.

Gribbin, J. (1988) 'The Greenhouse Effect – Inside Science No 13', *New Scientist* October 22nd

Hardin, G. (1968) 'The Tragedy of the Commons', *Science* 162, pp. 1243–1248

(IPCC) Intergovernmental Panel on Climate Change (1990) Policymakers Summary of the Scientific Assessment of Climate Change *Report Prepared for IPCC by Working Group I* Bracknell, United Kingdom: United Nations Environment Programme and the World Meteorological Organisation (June)

Karl, T., Baldwin, R., and Burgin, N. (1988) *Time Series of Regional Seasonal Averages of Maximum and Minimum Average Temperatures Across the USA* Ashville, North Carolina: National Oceanographic and Atmospheric Agency (March).

Kellogg, W.W. (1989) 'Carbon Dioxide and Climate Changes: Implications for Mankind's Future', in S. Fred Singer (ed.) *Global Climate*

Change: Human and Natural Influences New York: ICUS, pp. 37–65.

Landsberg, J.J. (1989) 'The Greenhouse Effect: Issues and Directions – An Assessment and Policy Position Statement by CSIRO' *Occasional Paper No. 4* Melbourne: Commonwealth Scientific and Industrial Research Organisation.

Marks, E.M., Swan, P.L., McLennan, P., Schodde, R., Dixon, P.B., and Johnson, D.T., (1989) 'The Feasibility and Implications for Australia of the Adoption of the Toronto Proposal for Carbon Dioxide Emissions', *Report to CRA Limited* (September).

Mason, B.J. (1989) 'The Greenhouse Effect', *Contemporary Physics*, 30(6): 417–432.

Newell, R., Hsiang, J. and Zhongxiang, W. (1989) 'Where's the Warming', *MIT Technology Review*, (Nov. and Dec.).

Nordhaus, W.D. (1990) *To Slow or Not to Slow: The Economics of the Greenhouse Effect* New Haven, CT: Yale University.

Pearce, F. (1989) 'Methane: the Hidden Greenhouse Gas' *New Scientist* May 6th.

Quesada A.U. (1989) 'Greenhouse Economics, Global Resources and The Political Economy of Global Change',*Environmental Policy and Law*, 19(5): 154–161.

Schneider, S.H. (1989) 'The Changing Climate' *Scientific American*, September.

Singer, S.F. (ed.) (1989) *Global Climate Change: Human and Natural Influences*, New York: ICUS.

Spencer, R.W. and Christy, J.R. (1990) 'Precise Monitoring of Global Temperature Trends from Satellites', *Science* 247: 1558–1562

The Treasury (1989) 'Developing Government Policy Responses to the Threat of the Greenhouse Effect', *Economic Round-Up – November 1989*, Canberra: Australian Government Publishing Service, pp.3–20.

Tolba M.K. (1990) 'Financing Global Environment Problems', *Address to Commission on Environment*, Document 210/333, United Nations, Paris 1990.

White, R.M. (1990) 'The Great Climate Debate', *Scientific American* 263(1): 18–25.

Appendix

National Priorities Project Sponsoring Organisations

Australian Chamber of Manufactures
Australian Chemical Industry Council
Australian Finance Conference Ltd.
Australian Goldmining Industry Council
Australian Institute of Petroleum
Australian Small Business Association
Australian Stock Exchange
Business Council of Australia
Pharmacy Guild of Australia

About Tasman Institute

Tasman Institute, established in 1990, brings together a network of scholars, business leaders and other associates. The Institute aims to make a significant contribution to advancing the level of understanding of the major issues involving governments, society and the economy.

Tasman Institute is supported by Australian and New Zealand business organisations, and is governed by a Board of Directors.

The Chairman of the Institute is Mr S. Baillieu Myer AC, Chairman of National Mutual Life Association of Australasia Ltd and Deputy Chairman of Coles Myer Ltd.

The Deputy Chairman is the Hon. Sir Roger Douglas, a former New Zealand Finance Minister.

The Executive Director of Tasman Institute is Dr Michael Porter, who was formerly the Director of the Centre of Policy Studies at Monash University.

Tasman Institute has a team of researchers and policy analysts of outstanding quality under the leadership of Dr Porter, and Director of Research, Dr Alan Moran.

Dr Moran was formerly Director of the Federal Government's Business Regulation Review Unit and First Assistant Commissioner of the Industry Commission. Dr Moran is the inaugural Director of the Institute's 'Markets and Environment' project.

Based in Melbourne, Tasman Institute is an independent economic and environmental policy research organisation. The Institute has no political affiliations.

Tasman Institute produces regular publications and reports, and hosts regular evening seminars featuring topical speakers of international standing.

Tasman Institute Pty Ltd & Tasman Economic Research Pty Ltd
(Incorporated in Victoria)
457 Elizabeth Street
MELBOURNE VIC 3000
AUSTRALIA
Tel:(03)326 8033   Fax:(03)326 8002